Re-Volt!
Our Impending Energy Revolution

By

Jeremy Gorman

Strategic Book Group

Copyright © 2011 Jeremy Gorman.

All rights reserved.

No part of this book may be reproduced or transmitted in any form or by any means, graphic, electronic, or mechanical, including photocopying, recording, taping, or by any information storage retrieval system, without the permission, in writing, of the publisher.

Strategic Book Group
P. O. Box 333
Durham CT 06422
www. StrategicBookClub. com

ISBN: 978-1-60976-742-6

Contents

Introduction iv
1. Energy is Everywhere 1
2. Energy Independence 9
3. Conservation 33
4. Geothermal energy 42
5. Wind in the Eaves 46
6. Solar Power 58
7. Nuclear Energy 66
8. Hijacked Hydrogen 75
9. Ossified Electricity 82
10. Wasteful Water 89
11. Tides 95
12. Ocean Currents 99
13. Combined Solutions 103
14. Do It Yourself Energy 107
15. Alternate Energy Proposals 115
16. Opportunity 122
17. Matching The Source To The Need 126
18. Wind and Water 129
19. Energy Alternatives 133
20. Power House 140
21. Transportation 145
22. True Alternatives 155
23. Alternate is Supplemental 159

Introduction

Energy is what makes things happen. It comes in many forms. Your internal energy is the most critical. It gets you up in the morning and lets you accomplish things. The rest of your energy you buy.

Energy not only comes in many forms, it is used in many forms. We need energy for heat, light, food and for moving things. We need energy for communication and calculation. Each of these demands different forms of energy, so each will be best served by a different source of energy. So far man has not been very clever about his choice of energy sources for his different needs. For example, heat is best served directly by the sun, or by stored solar energy in fuels. Wind, water, wave and tidal energy are not well adaptable to heat production. Communication and calculation are better served by photovoltaic, wind or wave energy because they use the low voltages supplied by photovoltaic and wind sources. Making electricity from coal and sending it 100 miles to your computer is a wasteful enterprise. We can easily replace that system with practical sources close to home. *Re-Volt!* is the manual for matching the source to the need. That is how we will make the energy of the future effective for everyone.

Man has been increasing his energy demand since he first lit a fire about 30,000 years ago. We buy ever increasing amounts of coal, oil, electricity, gas and wood. As a result, those supplies are decreasing and their cost is mounting. We have cut down over half of the world's forests.

Jeremy Gorman

We have used up all of the easily accessible oil that nature left for us. And that's not all. Burning those fuels is changing our atmosphere. We are making the earth warmer. We face a dilemma.

Energy is the world's largest industry, with the possible exception of food, and food itself is a huge energy consumer. In the next thirty years the world is going to stand that industry on its head, as the cover of this book shows. This revolution will affect more people than the Industrial Revolution (about 2 billion people) or the Internet Revolution (almost 5 billon people). You will be involved, whether you like it or not. So join the fun. This book is your survival guide. It discloses, explores and explains not only present practices, but also how we got here and what you must do to make the coming change beneficial. Better still, *Re-Volt!* invites your ideas and your solutions to the problems we face. The problem is immense, but there is no problem too big to solve if we all get involved. Join in. You will be involved anyhow, so it might as well be by choice and in support of your goals and aims. If we do it right, it will not only be informative, it will be fun. If we work together, we will all be winners--even Big Oil.

Energy is everywhere. We are bathed in more solar energy every day than all mankind uses. In addition we have the energy of wind, rivers, geothermal, tidal and ocean currents. We ignore 99.9% of it. Why? We have had the science to use the energy of the sun, the wind, the tides, the streams and the geothermal energy in the earth for over 100 years. We need to build an alternate energy industry to supplement the current system and take control from Big Energy and put it in the man-in-the-street. We need to stop warming the earth. We need to keep the price of energy affordable. We can!

From the World War I until about 1980 the United States was the manufacturing dynamo of the world. Then American management decided that our labor costs were too high and they began shipping our manufacturing facilities overseas to capitalize on cheap labor in countries that suppressed their people. Since then, they have shipped at least 13 million U.S. jobs overseas. Those jobs will not come back, because we also gave away the manufacturing know-how that allowed us to out-produce any other workers in the world... That leaves 13 million Americans unpaid or underpaid as they take menial or part time jobs to support their families. That means that our U.S. market has declined by at least 10%. The unrecognized problem is that the low-paid workers in the foreign countries still can't afford to buy what they make, so their output has to come back to the U.S. to be bought by a smaller buying public. We shrank the world market by underpaying its workers. The executives, meanwhile, have never had it so good. Their salaries have more than tripled in that time. They are bleeding their own workers. They love globalization, because if we can't support them in the U.S. , they will move their headquarters to Bali, or Shangri La. Already Exxon Mobil pays no U.S. income taxes on their $20 billion earnings because their earnings are either claimed to be in the Bahamas and Bermuda, or because they have finagled special tax relief for developing "New" sources of fuel. Their lowest paid U.S. worker pays more U.S. income tax than they do!

We have millions of creative unemployed people who want to work. Even a first grader can put those ideas together and ask "Dilemma? What dilemma? That's an opportunity!" Let's employ those millions in alternate energy instead of financing Al Qaeda and the Taliban with oil money. Let's use those new technologies to stop melting the polar ice cap. Let's make environmentally sound new products to sell to other needy nations and eliminate our outrageous trade deficit. Let's make small energy efficient cars that will sell anywhere in the world. Let's manufacture energy devices of value so our GDP increases and pays our monstrous national debt. Where *is* that dilemma?

Long overdue, and exacerbated by our "let the other guy do it" attitude, the vast world of alternate energy is finally beginning to come into its own here in the United States. It has already begun in many other countries. Burning stored solar energy resources is so easy that we have neglected other sources of energy. Most of our energy comes from the sun, which bathes this earth with six times as much energy every day as all mankind uses. We ignore, neglect and undervalue 99.9% of that immense resource as a matter of routine daily life. Wind, waterfalls, wood, and fossil fuels—these are natural sources of solar energy. Nuclear, tidal, ocean currents and geothermal energy are sources of non-

Jeremy Gorman

solar energy. We will use them all. Our insatiable demand will force us to. *Re-Volt!* explains all that in detail. It is your guide to the creation of an entire new alternate energy system to supplement the one we already have. It offers creative solutions to the use of all those alternatives. It also invites the reader's ideas. We are all going to be involved. Let's enjoy the process.

Many of us think that some great cosmic flash will come along and solve our energy problem. No Way! There will be hundreds, perhaps thousands of alternate energy sources that fill little energy niches formerly filled by oil, coal or gas. That is as it should be because energy, which is used everywhere, exists everywhere. Current supplies have been so easy, that we ignored an overwhelming supply, while complaining about rising fuel prices. The science for using these alternate sources has been around for over 100 years, but the technology to capitalize on it is retarded, leaving a growing gap between demand and supply. Oil, coal and natural gas are wonderful resources, but they are finite. We have used more oil than we have discovered for over thirty years, and our demand is increasing while the supply is decreasing. We don't know if we will run out of fossil fuels in 5 years or 500 years, but we know that we will!

What is this energy revolution? Many of us think that it will be the utilization of alternates to the burning of fossil fuels. They are half right. The real revolution will be the transfer of *control* of energy from a few profit-driven corporations to the man in the street. Note that it is not the transfer of all energy sources from those huge organizations, but *control* of energy. They will continue to supply 40% to 70% of our energy. That is as it should be. They have been extremely effective and creative. That is why they gained their position of dominance. However, dominance is an invitation to abuse. The massive energy industry has abused that dominance in the past, and will do so in the future, unless we make a concerted effort to wrest that control from them. Do you want to pay $8/gal for gasoline in 2016? You won't if you help establish alternatives that can be generated in smaller amounts almost anywhere. Rather than create a few massive alternate energy companies, we need to create thousands of small alternate energy companies that employ you and your neighbors. That will keep the energy companies in line. When OPEC arbitrarily raises prices again (and they will) customers can say, "Forget it! I'll make my own or get it someplace else."

Over 95% of our current energy supply is sun-derived. Oil, gas and coal are stored solar energy created over millions of years by living creatures. Green plants used solar energy to convert atmospheric carbon dioxide, the raw material or ore, into living things. At death a slow decaying process forms fossil fuels. When we burn those stored energy materials we put that carbon dioxide (CO_2) back into our atmosphere where it absorbs heat and causes global warming. Man has created an exponential increase in energy demand that has gotten out of control. That increase is destroying that delicate environmental balance that spawned essentially all present life forms. If our outrageous demand continues to be out of control, we will cause a major environmental change worldwide, like the one that destroyed the dinosaurs, the dominant earth species 60 million years ago. As the current dominant species, do we want to be next?

Re-Volt! has three major objectives:
1. Look at our current energy supply and examine its flaws and potentials.
2. Analyze and capitalize on our myriad alternate energy resources and develop a new system to supplement our supplies. We should develop an alternative source for every kind of the energy we use.
3. Restore the balance of our largest energy raw material, atmospheric CO_2, to the 280 ppm that spawned all current life forms on the planet.

Re-Volt! does not take a "good guy, bad guy" approach to our current energy source. We all are the bad guys. We have created an outlandish energy demand and didn't bother to look at its basics. Energy suppliers answered that call and did it well. We all made the mistake of looking for the easiest rather than the best sources of energy. Those sources were not the cause of our profligate and wasteful use of energy—we were. Energy suppliers just capitalized on our shortsightedness. It's time we faced up to our error. We may be forced to.

Chapter 1

Energy is Everywhere

Thanks to Al Gore's marvelous *An Inconvenient Truth* we are all finally aware that energy is a critical resource that is abused, misused and wasted. But there is more to Al's message, although it seems less obvious to most of us. Energy is *our* issue! It is not Al Gore's, or Uncle Sam's. It is not GM's, or Con Edison's. It is yours—and mine. Man has been quite creative about energy, and has developed ever better sources of energy. But it is important to note that we have without exception chosen the easiest route to energy--not the best. Prehistorically, we burned wood. Then along came coal and eventually oil. In each instance, we overemphasized the easy, and discounted the others.

An Inconvenient Truth revealed to all of us that the energy problem we face is not entirely new, and is much more complex than most of us believed. It comes in three forms. First our demand exceeds our current sources of supply. Second, that current supply is literally changing our earth's atmosphere, perhaps fatally. Third, the expense, not only financially, but politically, is out of control.

Let's take a look at demand. Our current fossil fuel usage (U.S. figures) is outrageous:

U.S. Energy Consumption

Source	U.S. Annual Usage	Energy Produced	CO_2 Emitted
Oil	1.215 Billon Tons	43.6 Quadrillion Btu*	3.74 Billion tons
Coal	1.128 Billion Tons	37.2 Quadrillion Btu*	3.90 Billion Tons
Nat. Gas	0.417 Billion Tons	19.7 Quadrillion Btu*	1.15 Billion Tons

* British Thermal Unit-heat required to heat a pound of water 1° F

That's 11 tons of fossil fuel each year for every man, woman and child in the United States. It also represents 8.3 Billion tons of CO_2 emitted (27 billion tons worldwide). Take note of a major difference between greenhouse gas (CO_2) emissions from our various energy sources. Coal produces 4774 Btu per pound of CO_2 emitted, while oil produces 5836 Btu (a 22% increase) and natural gas 8560 Btu per pound of CO_2 emitted (a 79% increase). Those are significant differences.

Important facts: 92% of that coal and 26% of that natural gas makes electricity. Because of other sources, including hydroelectric power (6%), nuclear power (19%), and wind power (1%), only 48% of our electrical energy comes from coal, our worst polluter. Note that electricity is not a source of energy, but a carrier of energy. It is our most popular method of moving energy from one place to another.

A word about the significance of the greenhouse gas emissions: The earth's atmosphere contained 1.4 trillion tons of CO_2 for at least 60 million years. All life on earth grew up in and adapted to that delicate balance. I call it a delicate balance, because life continually generates CO_2 as things die, decay or get burned and as animals breathe. But trees, plants and algae constantly take atmospheric CO_2 and create new living things out of it using the energy from sunlight. Starting about 700 years ago, man began cutting down the trees and burning fossil fuels in huge volumes. We screwed up that balance and, just in the past 100 years, have increased the atmospheric CO_2 by 540 billion tons—a 39% increase in one century. Do you really believe that the plants and animals that live in this atmosphere could completely ignore such a major change in a vital source of their livelihood? Carbon dioxide forms an acid in water and in the blood. There is a remarkably sensitive balance in your system. As the blood CO_2 increases (called hypercapnia), the acid increases and your blood vessels dilate to allow more blood to flow and supply more oxygen and flush out the extra CO_2. Changes of ten parts in a million will change the acidity enough to cause that process to begin. Man has increased the CO_2 of the air you breathe by 100 parts per million in the past century. It seems probable that such a change will change the chemistry of your body substantially. Are we about to cause serious physiological changes in people and animals all over the world?

We are incredibly wasteful of energy. We waste as much as we use. (Yes, that includes you.) We could reduce our energy usage by 20% without missing a single TV show, a single meal, a single ride to work, or causing our homes to get uncomfortably cold. Energy has been so cheap that we didn't really consider its cost.

By far the easiest way to obtain energy is to burn something. Nature has been storing solar energy for billions of years, and it is easy just to light a fire and live off that stored energy. There was plenty of it when the world contained half a billion people. So mankind was casual and profligate in its use. Each of us demanded more and more. When you build a fire in an open pit, fully 95% of the heat you produce is wasted and just heats and pollutes the atmosphere. However, the forest was full of trees, so we just cut down some more.

We got a few warnings long ago that this was not the best way to get your energy. Easter Island completely denuded its environment, and the culture just disappeared somewhere between 700 and 1700 AD. It had consumed all its natural energy resources and could no longer sustain itself.

We got another warning in Greenland. The Norse established a culture on Greenland in 985 AD. It had essentially the same fate in 1425 AD as had Easter Island. In that instance, however, they went back to Norway.

We didn't learn. We continued on the easy route--burn up all that stored energy, and the devil take the hindmost. The problem is that the devil will! And it looks like, after 6000 years, we will be the hindmost!

Americans began considering alternate sources of energy somewhere around the Civil War. A few "nutcases" suggested that we had better look for other energy sources. But coal and later oil were so easy, and there was so much of it, that those oddballs were disregarded and even mocked. Despite this, however, windmills were a major source of power on many farms for over 100 years. Windmills were mostly used to pump water, or mill grain, but they were a substantial source of renewable energy. In 1859 oil was discovered in Pennsylvania. That began a major change that ultimately completely dominated the energy industry—even dwarfing the use of coal, which had been used for over 800 years. There is no doubt that oil was the most cost effective energy source and so it dominated the energy industry worldwide.

In 1859, there were less than 2 billion people on earth. Many of them lived in warm climates that didn't need much energy, at least not for heat. Mankind went crazy building coal and oil-fired plants everywhere. Efficiency? Pollution? Who cares? We have an unlimited source of energy. Why not use it?

The problem was that it was not unlimited and pretty soon we were looking for oil all over the world. Demand grew exponentially, but supply did not! Enter international politics. Some countries in which Americans had discovered oil decided that they had an opportunity to capitalize on our outrageous demand for oil. Why not nationalize the oil wells and charge more for the oil. The Organization of Petroleum Exporting Countries (OPEC) was formed by five nations in the 1960's. They began nationalizing oil supplies and grew to 13 nations by 1972. They agreed to an oil embargo in 1973 for the purpose of raising the price of their oil. By then they were a major source of worldwide oil. Since OPEC's embargo we, as a nation, have made some improvements in our energy efficiency, but we still have a long way to, go. In fact, despite the embargo, our U.S. imports rose from 45% of our oil consumption in 1973 to 65% in 2006. Meanwhile, OPEC representatives meet regularly and set the price of oil arbitrarily to their advantage.

Efficiency is not the whole story. We just turn up the heat or drive ten miles to the shopping center. That must stop. We have to learn about energy conservation—a concept we neglected because energy has been so cheap. That will take a massive public education program. It will also find lots of opposition among the people it affects.

The OPEC embargo gave us fair warning in 1973, and we listened briefly. But as soon as the price came down again, we abandoned our efforts at alternate energy. How could we have been so, stupid? They had already showed us what their intentions were, but we didn't want to listen. We threw away 35 years.

So it happened again in 2006. The price of oil began increasing, slowly at first, but increasing rapidly in a short while. But we had given away our opportunity, and we were paying $700 billion per year for oil from antagonistic foreign governments. We had not developed the alternatives we began in the 1970's, and had become even more dependent upon oil. The oil companies loved it. OPEC loved it. They thrived on it. They exercised political power and disrupted trade. They invaded our financial system. They challenged our international efforts and called us bad names. In fact they ruined our worldwide reputation. And we let them!

Can we learn this time? Of course we can, but remember; this is our task-not our government's or our oil companies. When gasoline was $4.50/gal, and many people couldn't afford it, altruistic Exxon Mobil recorded the largest corporate profits in the history of mankind. There is the message! The task is ours---it is not the governments, and it certainly is not Exxon Mobil's. It is not even Al Gore's. We must seize the sources of energy and bring them back to the people. That is the goal of *Re-Volt!* I hope to be able to stimulate every one of us to take it seriously. This will be your Re-Volt! It will not be hard. Energy is in your back yard already. All you need to do is learn how to use it.

There is another facet to this. Nature has distributed energy everywhere. Sunlight falls all over the earth. Water is pretty largely universal—so much so that we call water-deprived places deserts. Forests grow on every continent except Antarctica. However, we get almost all of our energy from small pockets of energy rich materials. Most of the energy that you use comes from hundreds of miles away, while at the same time we ignore the energy that is in our back yard.

Isn't it about time we took a close look at that process? Is this really the best way for us to get powered up? Why ship oil from Saudi Arabia to New England when energy abounds in New England? We have gotten off track. Actually the commercial interests that supply energy have led us off track. That must change, but it won't happen by itself. Big Oil will oppose all such effort and they will use some of their profits to under inform or even misinform you to make sure that change does not occur. Shall we let them?

Don't get me wrong. Big oil is not a bad guy! He is a commercial guy whose primary purpose is to make money. The better he is at supplying our energy the more money he makes. Who can blame him for advertising what he has to sell? It is *our* job to find other sources. Big oil will not unless they can make money on it. They already own much of the oil sands in Canada and much of the oil shale in the American west.

Aaaahh. There's the rub! Oil and coal companies don't own the shale oil, or the oil sands. In fact they don't own the oil in their own oil fields, or the coal in their mines. What they do own is the right to mine and exploit them. Those are natural resources. They belong to the earth. We have had some confusion about that for about 100 years, since the oil companies convinced our government to give them a depletion allowance for the oil they drill. They are depleting a natural resource and should instead be charged a depletion tax for the declining natural resources on which they have a license to make a profit. That is one reason oil has been such a bargain, and a major reason for our profligate use of it.

Big Oil didn't create that oil. Nature developed a method of storing solar energy. With the intercession of chlorophyll, living plants use some of the energy in sunlight to convert water and carbon dioxide from the air into sugars, starches, cellulose and other organic tissues from which living creatures are assembled. When they die, they are processed by trillions of microbes into "fossil fuels"—the name we give to organic matter that can be burned to release the solar energy that they have absorbed. The coal, the oil, and the natural gas were made out of the carbon dioxide from our atmosphere. Burning them returns that carbon dioxide to that atmosphere. That seems like a normal cycle. The least we can do is to make good use of that energy we extract in that burning process. We don't. And worse, we continually destroy the very trees that make that conversion for us. Remember Easter Island?

Man gets carried away with his projects. We are burning hundreds of times as much organic matter as nature is creating. That is at least in part because we keep cutting down nature's trees that clean our atmosphere for us. The net vegetation content of the earth is declining, because we are so profligate in our use of energy. In the past energy has been so cheap, that we didn't bother to be careful or efficient in its use. We waste more than we use. We are profligate in our reproductive process as well, and the human population of the earth has exploded. As energy became more available, we created more humans to use it. We overdid it. We are returning far more carbon dioxide to the atmosphere than the remaining plants are consuming. The carbon dioxide content of the atmosphere was about 280 parts per million for at least a million years--probably much longer. In one century, man has changed that. We brought the carbon dioxide from 280 to 390 parts per million in one century. In doing that, we discovered something. Carbon dioxide is a sunlight absorber. As it increases, things get warmer. The carbon dioxide retains more of that solar heat. For at least one hundred million years nature, which is remarkably adaptable and flexible, built our current ecosystem. And generations of its inhabitants adapted to that system. Suddenly, at least on a geological scale, we have changed that ecosystem. It already has caused remarkable changes. We are melting the polar ice caps, which apparently had a much greater influence on our climate than we realized. Whole species are dying, and others are moving. Ocean levels are beginning to rise, and to threaten low-lying coastal areas. Worldwide about one hundred million people live within 25 feet of sea level. A 20-foot rise in the oceans would cause untold human displacement and misery.

Some claim that these are normal changes caused by nature. Please note, however. We had a major climate change once before. It completely wiped out the dominant creatures of this earth--the dinosaurs. As the dominant species on earth today, we ought to at least consider that. Are we next? Suppose we aren't the cause of the climate change. Working to stop it certainly won't hurt. It might employ thousands of our currently unemployed. If we are a contributing factor, such action may well save millions of lives and perhaps the whole race of man. Considering our past history I'm not sure that man's extinction would be such a bad thing for this earth.

We are now in a serious recession of our own making. This book contains many ideas for helping us out of that recession. These are not the only ideas. They may not be the best ideas. They are largely confined to energy production and use. Most important, they are stimuli for your ideas—the ones that will employ us, make us energy independent and pollution free. The ones that will bypasses the carbon cycle and change us from a burning society to a society that goes directly to the sun for much of our energy. One that bypasses Big Oil, The Grid and Enron. It can be done. It will be done *if* we all join in and do our part every day. It is a fertile field. It is a sponge for ideas—your ideas. We literally have a world to save. The caveat is not "Yes they can", or "Yes you can", but "Yes *we* can!"

It may be worthy of note. The characteristic of man that has set him apart from all other creatures is his ability to cooperate. Almost everything we see and use in life today is a result of the cooperation of man. Buildings, roads, automobiles, computers, cell phones---even grocery stores and shopping malls, are the result of our working together. Why do we spend so much of our time fighting? Getting together to revamp our energy supply will require cooperation. Look at this, though. Cooperation is fun. We meet wonderful new people and learn fascinating new things together. Have you a better idea? Let's hear it.

While addressing cooperation, please take note. At the end of every war we always have a negotiating session to settle the terms we fought over. Grant and Lee at Appomattox was a famous example. Why don't we negotiate first, and eliminate the war part of that equation? We continually say that we don't want to talk to Iran or to North Korea. Why not? Isn't that better than fighting with them? *Re-Volt!* is about cooperation. Let's eliminate the fight.

Chapter 2

Energy Independence

President Bush said in his 2006 State of the Union speech that we are addicted to oil. In fact we are even more addicted to energy consumption. Oil is the easiest source of energy, and the cheapest--*today*! But that era is coming to a close. Oil is getting scarcer, harder to extract and more difficult to process. The price goes up and will continue to go up. Our government has taken few steps to minimize our dependence or to seek alternative sources of energy. In fact, it continues to provide grants and subsidies to our most profitable industry while simultaneously forbidding them to drill for oil in many areas. There are good environmental reasons for those bans, but that doesn't improve the world-wide energy supply.

Many of us seem to think that some brilliant solution will appear like sunrise to solve this problem. In reality however, there already are many solutions each waiting to fill a niche need. This is a highly desirable outcome, because it moves control of energy sources from a few big monopolies to a diverse group of competitive suppliers. When the price of gasoline goes up today, you have no choice but to pay it. If there were many alternatives, you could opt for another source. Better still, if there were alternatives to driving your car that would make a bigger difference.

We are faced with three major energy problems. Although related, they are not the same. The first is that we are dependent upon fossil fuels for over 90% of our energy today. (About 8% of our electricity comes from nuclear power, and another 8% from hydroelectric power.) Many oil sources are from unstable foreign governments that would like to manipulate the price and supply not only to squeeze as much money as possible out of this wealthy nation, but also for political purposes. Oil is, and will increasingly become, a political tool. Venezuela, Saudi Arabia, Iran and Russia already wield this power ruthlessly. It will get worse. That is why we went into Iraq.

The second problem is that our two largest energy sources, oil and coal, are serious greenhouse gas emitters. We are creating global warming that is already beginning to cause serious long-term problems worldwide. When we seek solutions, we need to face both of these problems, but not necessarily with the same urgency. Ultimately, we must develop renewable energy to cease or greatly reduce greenhouse gas emissions. We must keep both problems in mind as we seek solutions.

The third problem is not so obvious nor as well recognized. Essentially all of our energy comes from a relatively few large profit-oriented companies. They actually control our energy--its supply, its cost and its usage. The biggest change in energy that is coming will be the removal of that control from the big corporation to the user and a host of small competitive suppliers. Today we have few choices for energy. You can go to The Grid, to Big Oil, or to Coal. Oh, yes, a few of us can go to wood, but that too is a declining source. If we stay on track, the control will become ever more rigid, and the price will keep escalating. When we calculate the costs of making the changes, the factor that we can't figure is the increase in cost of existing energy sources. As that remains in the control of a few massive corporations and some foreign countries that want to squeeze us for money and power, the costs will explode. *That* is the problem we must avert!

We are already fighting over energy and global warming. There will be naysayers in any major change in a democratic society. That is the nature and the strength of a democracy—the ability to be heard. In this instance we need to look at a little realism. Global warming is a fact. Even the naysayers can't change the fact that each of the last three decades has been the warmest on record. Even the naysayers can't put ice caps back in the arctic where they have been for millennia. They don't even try! There is much evidence that man is a major contributor to global warming. Evidence is not proof, but as it mounts it becomes proof. So far the naysayers don't present any evidence that man is *not* a major factor in global warming. They dispute the facts presented by the scientists, but offer no contrary evidence.

The naysayers should consider this. Either Man is a contributing factor to global warming or he isn't. If he is and we take action now we can perhaps gain control of global warming. It will cost a lot of money, but we can make a real international effort and try to eliminate man as a contributing factor. If Man is a cause, we can mitigate that problem and employ millions of un- or under-employed people doing it. If Man is not, then the world will dry up and we will all be forced into a complete change of lifestyle, or just fade away and become an extinct species like the once-dominant dinosaurs that perished in the last major climate change. The money we spent will no longer matter, and we will have employed millions of people in the interim. If Man *is* a factor in global warming and we don't take action, what will you tell your starving grandchildren? That you wanted to save money? Or get re-elected? Look in the mirror and ask yourself that question.

These are not the only reasons. Fossil fuels are a finite resource, and we will eventually run out. People don't want to believe that, but finite is finite. We are attacking that supply with ever-increasing vigor. We already are surprised by how much we have drained that resource. We have less than we claim we do. Oil reserves in Saudi Arabia, Iraq and Iran are seriously overstated. We keep searching ever deeper and ever more distantly for more fossil fuels, and the pressure to invade wilderness and wildlife refuges increases with every year that we remain fully dependent upon fossil fuels. The question is not *if* we will run out of fossil sources, but *when.* Without some serious effort from mankind, that will not occur until we have put all that carbon back into our atmosphere and our oceans and made our planet as warm and as lifeless as Venus. Renewable resources will not disappear. We will not run out of sunlight, wind, rivers, geothermal heat or tides.

How did we get into this mess? It is pretty straightforward. Until about 150 years ago, we had little advanced technology for making use of the world's energy. We burned wood, and then coal and finally oil—

all sources of stored solar energy from nature. These were not always easy for individuals to get. Not everyone had the tools to cut down trees, to mine peat or coal, or to drill for oil. We employed large businesses to do that for us. They did it very well. So well, that we have, in 150 years, built a society that is intimately involved with fossil fuels as a foundation for all the energy we use. Our machinery, our farms, our recreation and our transportation have all been built upon the easy access to fossil fuels. We don't want to face the fact that it is not an unlimited supply--not only for its energy, but for its other products as well. Our plastic industry, which makes our shopping bags, our computers, our telephones and much of our clothing, is also built upon oil as its raw material. It will be necessary to change, or at least modify that too. Our population, and therefore our demand, continues to grow. If that keeps up, there will come a time when no amount of energy will satisfy our needs.

It will not be easy, and it will not be cheap. It will run into extreme opposition from those who like the world as it is today, and will look at these changes as frivolous invasions of their utopian world. But facts are facts. We must change. The sooner we do the less disruption it will cause. At least as much of the strain will come from changing our cultural adaptation to oil as will come from the conservation and replacement of it. That will be made harder by the fact that we have grown dependent upon oil not out of whim or fancy, but because it is our most cost effective source of energy. We will pay more for energy as time goes on--not just in dollars, but in its percentage of our cost of living. The more effort we make now to modify the technological and societal problems we face, the more successful and the more easily we will make the transition. But the transition will come—like it or not. Let's act now to make the transition as smoothly and as economically as we can. Using our creativity may make the transition manageable, but there will be no cheap solution!

As our demand for energy increased the energy companies got larger and more powerful. Note that the largest annual profit in all history was by Exxon Mobil—an energy company. They performed a vital function to our expanding culture and its increasing energy demands. Their techniques became ever more sophisticated and involved—and more expensive too. So the energy industries grew and prospered, as our energy profligate culture demanded ever more. Soon they developed a dominant position in our culture and, of course, some of them began to abuse it. Enron is the common example of that abuse, but they were neither the first nor the worst—just the most blatant.

Abuse is not the whole story. The overwhelming problem was increased demand. We not only used more energy as the population grew, but we also increased the demand per person. Partly because of the ingenuity of the energy companies, energy remained cheap. Not only was the supply limited, it was unevenly distributed, so the transfer of energy from diverse sources became ever more complex. Demand continued to grow and the energy companies made ever bigger and more expensive facilities for supplying it. As they became bigger and more sophisticated, they became more efficient as well.

We were developing a pattern of making huge investments to supply huge amounts of energy to the people. There was a flaw in that pathway. Although we made bigger and more expensive equipment for our power demand, energy is not used only near power plants. It is used everywhere. So we kept shipping it farther and farther from our ever bigger energy sources. However, energy is available everywhere. Our power companies made their energy so easy that we completely neglected the myriad other sources of energy that surround us. Sun, wind, water and geothermal energy engulf us daily, but we ignore them and continue to buy our energy from the big sophisticated energy companies. With the help and encouragement of Big Energy, we established a culture of wasting energy from the easiest sources. While the energy supply was large and cheap, that was OK.

As you would anticipate in such a circumstance, politics became involved. OPEC was formed specifically to extract more money from the energy-profligate West. It is a sad fact that they were right---we had grossly underpriced oil in the United States and lived on that false economy for decades. (Recall the depletion allowance instead of a depletion tax.) That is a major contributing factor in our extreme expansion of energy demand, not only as a growing population, but for every individual within it. And that was only step one. The politics of energy became a worldwide hassle. It will continue to do so as long as we remain dependent upon a few randomly located prime sources of energy.

Technology to the rescue! In the past 100 years we have devised myriad energy extracting devices that do not require massive publicly subsidized power facilities. They can and should be located at or near the site of the demand. That is the energy revolution that we are finally facing today. Long overdue and still sadly neglected, those local sources are the key to our energy future. Not every energy extracting technology requires massive installations. Many do, but those that don't should be everywhere. Because of the dominance of Big Energy, the local sources have been neglected. We must escape the oil cartels and the energy consortiums that control our use of energy today. As long as we remain dependent upon a few massive sources of energy, we will remain subject to their seizure by despots who want to squeeze us for all they can get. Ahmadinejad, Chavez and OPEC will have successors as long as we neglect the energy resources that they cannot control.

Oil is our dominant source of energy because it is the most adaptable, cost-efficient and available source. There will be no single replacement for oil. There will be hundreds, each adapted for a particular niche, or set of niches. If they are to be long-term solutions, they must be renewable. That means wind, solar, geothermal, rivers and streams or tidal sources. There is an abundance of solar energy, but it comes in many forms, none of which is extensively utilized today. Wind and rivers are two major underutilized sources of solar energy. Solar heating and photovoltaic panels that convert sunlight directly into electricity are also grossly underutilized. That is partly because they still are grossly overpriced. They are not mass produced with efficiently designed production systems. They are viewed as luxury items to prove that you are environmentally conscious. As the price of oil continues to rise each of these will become ever more economically viable sources. We must also make serious effort to bring down the cost of those alternatives, so that everyone can use them, not just the affluent. That is where your individual creativity will become a major asset. We have established a culture in which alternate energy is not taken seriously. They are primarily expensive showpieces. Not every new technology needs to be run by a public utility. Many technologies are better on your roof or in your back yard. Windmills, solar panels, photovoltaic panels and geothermals are all suited to small local installations. (Geothermal, ocean currents and tidal resources are more practical in midsized community installations.)So far these have been woefully neglected primarily because of the creativity of those public utilities. Education about alternate energy is a critical need today. It isn't even in most of our schools or industries. Most households haven't a clue about how to find and use alternate energy. How many of you have a true understanding of alternate energy? Information is readily available on the Internet, in this book and in your library or bookstore.

Windmills powered the farms of the Midwest and the west for over 100 years. Solar panels have been around for over eighty years. There is a small resurgence of these alternatives, but they will not become practical unless and until we all push to make it happen. Big businesses that now control access to energy oppose such developments because many alternative sources would wrest control of energy from them. We already suffer from over-centralization of energy sources. When electric power costs double, we pay it because we have no choice. Remember Enron? How do you battle the only supplier of a critical resource? Those commercial interests lobby our government for special treatment, and form PAC's to elect favorable government officials.

Government has been unwilling to put control in the hands of the people who pay for it. It will take a massive effort by *all* of us to bring about the changes we need for our future energy needs. What can *you* do to regain energy independence? It is not enough to complain and to point fingers or even write your Congressman. It is essential that each of us does something if we really want to solve the energy problems we face.

The largest and easiest part of the solution is to reduce our energy demands. Conservation not only works right away, but it doesn't cost anything—it saves you money. This approach addresses both the global warming issue and the single source issue. Our automotive industry builds 360 horsepower engines driving massive vehicles at outrageous speeds. They want you to build a home on four wheels that can go from 0 to 60 in 4 seconds. What for? Where? Two hundred sixty of those horsepower are completely unnecessary. Equating bigger with better is an American quirk. Why can't we build a car that weighs only three or five times its driver's weight, instead of fifteen or thirty times as much? We have become so dependent upon the automobile for transportation that the American car is no longer built as a transportation device. It is a status symbol. The auto industry says that they are building the cars that Americans want. Have you seen their ads? Which of you wants to put your family car into a 360 degree skid in a wet school parking lot or along a cliff at the edge of the Grand Canyon?

Most luxury things in a large car can easily be put in a small car. They already do that in Europe. Luxury cars should be the Jewel, the Diamond, the Gem or the Sapphire-small but extremely high quality. The Hummer and the Escalade should disappear. These ego-driven examples of conspicuous consumption are detrimental to all society and to our future. Even today our American automobile companies, rescued by the American taxpayer, still advertise all the luxury perks that are included in their new models. U.S. auto makers still believe in the "Small car, small profit" concept in automobiles. Meanwhile India makes the small, efficient and inexpensive Tata for Indians--the fastest growing auto market in the world. The American companies didn't even consider an export market. They had no interest in increasing the market, just in squeezing more out of the one they were already serving. Is that why we have such a miserable balance of trade? In fact they don't really want to serve the market so much as define it. Let some other country (Japan?) make cars for those 13 million Americans whose jobs were sent overseas.

How does one achieve this change? Primarily by education. This is such a massive problem that it will take all of us to conquer. Yes. That includes you. Few Americans have even a clue about how to conserve energy. It doesn't even enter their consciousness. That can also be guided by national tax policy. For example, the tax on new vehicles should increase as their emissions increase. Some states already have considered an emissions tax. We should also enact a horsepower tax—a $50 tax on every additional horsepower above 100 in a new passenger vehicle. That would put a $13,000 tax on a 360 HP vehicle. Even if that 360 HP is modified to run on non-polluting hydrogen, it still places a huge and unnecessary energy demand that will strain our energy sources. Such a hydrogen-modified vehicle could escape the pollution tax, but not the horsepower tax. The customer still has a choice, and it is up to him to decide. That tax money should be reserved for developing alternate sources of energy, or perhaps in educating the public on energy conservation. It should not go in 100 million dollar packages to Big Oil. They can already afford to spend research dollars for products that will keep them in control of energy. That tax money should go to Joe Shopowner who wants to build a windmill for his roof (and perhaps also that of his neighbors) that will cost $300, instead of $3,000. Energy tax money should go to small mass produced units which everyone can use instead of huge 2.5 megawatt windmills that scar the landscape, take 15 years and massive subsidies to build and feed electricity onto the grid where they can charge you a premium for it. (My electric company claims that their massive windmills will pay for themselves in four years. Nevertheless, they offer a special "renewable energy only" option to those who want to pay a 15% premium for their electricity. They have not explained how they can separate the two sources of electricity in my power lines.)

Another source, destroyed by a coalition of auto, rubber and oil companies in the 1930's, is public transportation--the streetcar! That requires a slow rebuilding process and lots of capital. Today it more likely would be a bus line, or an electric-powered tram. Why does Europe have high-speed rail service when we have larger distances to cover? Of course, high speed rail requires high population density, and will not be cost effective in rural Vermont. It could, however, be a major means of transportation along the Northeast Corridor from Washington to Boston and the California coast between San Diego and San Francisco. A line from New York through Buffalo, Cleveland and Detroit to Chicago might be cost effective. Up until about 1950, the northeast corridor supported a huge ridership on the local railroads, but the automobile became such a favorite that the ridership fell drastically and the railroads became obsolete and run-down at least partly because we built superhighways right beside railroad lines. It will not revive until we make public transportation less expensive and more attractive than the private automobile. In high density populations that is not only possible but desirable. Imagine New York City without the subway!

Then, there's the bicycle. Short intra-urban trips could easily be made on a bicycle, which is non-polluting, and might help solve our national obesity problem. You should demand the addition of bicycle lanes on your city streets. You might even buy a bike. For those who live in hilly cities like Cincinnati, consider a moped with a small 1 horsepower engine. (I was recently astonished to find that many so called Mopeds don't even have pedals.) In flat cities between Cleveland and Denver, a bicycle should become a primary means of home-to-work transportation. That is a cultural mindset which we might encourage if we want to conquer our obesity problem.

Alternative sources of energy must ultimately rely on less limited supplies of energy:the sun, the tides, ocean currents, wind, waterfalls, the oceanic thermocline and geothermal energy. The science to capitalize on each of these is already known, but the technology to bring it into practical use is lagging far behind. The faster we raid all our finite sources of stored energy, the more we hasten the arrival of a day of reckoning. The information technology revolution has expanded the industrial revolution to countries and areas that it had bypassed for over 100 years. Indians and Chinese are joining the industrialized world in record numbers, and are exploding their demand for energy. So far it has been primarily coal derived. China has lots of coal. This demand focuses upon the standard sources of energy because they are easier and more highly developed. Because of that China passed the United States in greenhouse gas emissions in 2009. However, China is learning as well. They also surpassed the United States on wind-generated electricity. They, too, have followed the massive windmill route, neglecting the small light-wind generator for the home. We may have an opportunity to help them divert their energy demand with technological advances that exploit other energy resources. We can export that technology and reduce our balance of trade deficit. But we have been lulled into a sense of comfort in the last three decades, and have almost completely neglected those other energy sources, which will become our mainstay in this century. Opportunity is staring us in the face. Are we finally beginning to see it?

The immediate problem is oil. We increasingly demand oil from areas in the world with unstable governments. Closer to home, we increasingly dig deeper and more risky wells to try to keep up with exploding demand. We have witnessed the BP oil spill in the Gulf of Mexico. That is just the beginning. The probability that some despot will come to power and try to hijack foreign oil supplies for his own purpose is high and increasing. There will be more Mamoud Ahmadinejads and Hugo Chavezes. Coal, a temporary solution to the oil dilemma, is more abundant than oil, but is more difficult to use. It also is in abundance right here in the United States, as was oil 100 years ago. Coal is a serious pollution threat. It has a larger percentage of carbon than oil. It emits substantially more greenhouse gases because it also has less heat per pound than oil. It contains more pollutants like sulfur, arsenic and mercury. Since it tends to burn hotter than oil, it produces more nitrogen oxides. The technology to minimize those emissions has been known for over 80 years. Industry touts the use of "carbon sequestration", a process that captures the CO_2 and compresses it for other uses. In over fifty years of such "Clean Coal" advertising, there is not a single carbon sequestering commercial enterprise in the United States. They are lying!

Coal, like oil, is a finite resource--it can and will be depleted. Coal cannot be used everywhere. Your hot water heater, your stove, your automobile and your clothes drier would be extremely difficult and expensive to convert to coal. (Perhaps you could find an old coal stove.) There are other choices. Many appliances are already electric powered --a wasteful and inefficient response. Many of us think that an electric stove is non-polluting. Actually, it is more so, but the pollution occurs at the power station instead of at your house. (I have already noted that 48% of your electricity comes from coal.) Why not convert boilers, gas stoves, driers and hot water heaters to hydrogen? That can be a completely non-polluting fuel that could be used in any of those gas appliances with a minor change costing less than $20. Automobiles can run on hydrogen. This, too, would take some of the monopolistic control of energy from the few massive oil companies to many independent suppliers from more diverse sources. Automotive companies, in an honest effort to increase the efficiency of a hydrogen vehicle, have moved toward the fuel cell. That is OK, but the fuel cell is much more expensive than the internal combustion engine—so much so, that fuel cell cars aren't really competitive in the market place---again their payback period is far too long. I might also note that the fuel cell is not adapted to macho overpowered cars that dominate many automobile advertisements. America has lost sight of the fact that automobiles are foremost a method of transportation. Their primary selling point is status. A fancy car is the ultimate showpiece of status. It would be better to use the less efficient internal combustion engine that can be adapted to operate on hydrogen with very little cost. Better yet, don't use hydrogen for transportation. Hydrogen is very difficult to store and takes up lots of space. To use it in transportation is difficult at best because you must carry a large, heavy container with you. Somehow, transportation has absorbed most of the effort on hydrogen because it would replace our largest single consumer of oil. People have not thought this problem through and get lost on tangents. The need is to have a universal non-polluting fuel. Hydrogen can be one such fuel.

Hydrogen can be manufactured by wind and/or solar-powered electrolysis of ocean water. It can be generated by power from ocean currents, which contain about 10 times as much energy as wind. The mass of the earth's oceans is 300 times the mass of the earth's atmosphere. Water is 800 times as dense as air and can drive turbines more effectively. In addition, ocean currents flow steadily all the time in well-known patterns, unlike the winds. Still further, if we create water driven power sources for the ocean currents, they can also be used in our rivers and streams. We don't have to build Hoover Dam to capitalize on the power of flowing water. We probably would have done better to build 30,000 small water turbines along the Colorado River rather than Hoover dam. Most of our nation is over 200 miles from an ocean current, but almost everyone lives within 100 miles of a river, and within 10 miles of a stream. Wind and water are truly unlimited resources capitalizing on an available but underused source of energy--the sun! Why do we ignore them? It seems to be largely because of a cultural mindset--energy comes from big companies, not from my roof or my back yard. And therein lies the immense pressure on oil companies to dig ever deeper and ever more dangerously for ever more oil. Consider the gulf coast oil spill. Perhaps you can develop a small tidal generator and make a business of it.

Strictly speaking, hydrogen is not a source of energy because it is not naturally occurring. It must be man-made. It is a superb medium for the transport and delivery of energy. Currently, hydrogen is made commercially in the U.S. by 9 different industrial processes. The largest of those (94%) is from methane--a carbon-based gas. Producing hydrogen from methane makes 6 times as much carbon dioxide as it does hydrogen. It merely transfers the source of pollution from the user to the manufacturer and it still depletes our stored solar energy reserves. Where's the long-term gain for anyone except Big Oil? At least one of those nine sources doesn't require fossil fuels--hydrolysis of water. (The reaction of steam on iron and the dissociation of ammonia are not direct contributors to greenhouse gases, but each requires high temperatures that need either fossil fuels or electricity to achieve.) Hydrolysis is currently the smallest source of commercial hydrogen. When we make hydrogen from renewable sources we progress toward independence from oil and limited stored energy sources. We also reduce global warming at the same time. Wind farms producing electrolytic hydrogen instead of electricity could help close the gap. It would be another step toward escaping the monopoly of "The Grid."In addition, the large wind farms should be out to sea where they do not cause NIMBY complaints from neighbors. Land need not be fought over and purchased. The wind there is faster, steadier, and blows much closer to the surface than it does over vegetated land. The towers need not be so high, and several levels can be stacked on each platform. The problem of wires and cables would be eliminated as well as the losses and extra expense incurred in phasing to the alternating current of the power grid. Best of all, it will disperse rather than concentrate the sources of energy. No more monopolies. A wind and/or ocean current driven hydrolysis station at sea would cost less than an oil-drilling rig. However, all the news is not good. Electrolytic production of hydrogen is only about 60% efficient. That is at least partly because you also create oxygen. However, oxygen is a major commercial product and can be sold as well. Hydrogen is extremely difficult to store and ship. Isn't this an opportunity for some creative problem solving? (See the chapter on hydrogen.)

To carry us through the transition period we could go to the oil sands of Canada and the shale oils of Colorado and the Rocky Mountains. These oil sources are more difficult to extract, but there is a huge supply of them, and they are out of OPEC control. They could produce an enormous cash flow for Canada, and reduce our dependence upon oil from politically unstable governments. The technology to recover them already is in commercial use. They are far more effective than drilling in the Arctic National Wildlife Reserve (ANWR) but more expensive. They are more expensive than oil or coal, but with the recent increases in oil prices are becoming more nearly cost competitive. They will *not* solve the greenhouse gas problem. Nor will they break big oil's stranglehold on energy sources, since big oil and coal interests already own the rights to much of the oil shale and oil sands. But beware! Like almost every mining process and other methods of using environmental sources, the oil sands industry is creating in Alberta a massive wasteland out of what was once a primeval forest. Man has a history if destroying nature in order to gain its benefits. We have the opportunity to stop this destruction, but we haven't even tried so far. We are creating a wasteland twice the size of Manhattan in Alberta--unnecessarily!

Electricity is supplied largely through coal. Although some power plants are reasonably good at controlling pollution they all could and should be far more so. The technology and equipment to do so have been available for 50 years. Greed and concern for the next quarter's bottom line have delayed their introduction. They constantly lobby congress for special consideration.

There are other options—nuclear, for example. That is a frightening concept for most of us---primarily because of Chernobyl, Three Mile

Island and ignorance. America's only big disaster in nuclear energy at Three Mile Island injured no one, although it frightened millions of people. Three Mile Island got out of hand because of inadequate training and no preparation for a failure. Realistically, they became overconfident. Even with Chernobyl, nuclear power has a far better safety record than either oil or coal. (See the chapter on Nuclear energy.)

Nuclear waste is a soluble problem. It is also a political one. Consequently we haven't done much yet. High-level nuclear wastes contain fissionable materials that can be reprocessed into nuclear fuel. Low level wastes have no fissionable materials, but have half-lives of 100 years or less. (A half-life is the time it takes a radioactive material to radiate away half of its radioactivity.) High level wastes can be reprocessed into new fuel rods for nuclear power plants. Low level wastes are a serious radiation hazard that is decreased by about 90% within two years. After that, we can dilute the low level waste with the spent ores from which they came, and return it along with some young trees to the mines where they came from as part of a mine reclamation plan.

Another advantage to nuclear power is that it can consume weapons grade material... What can be concentrated can also be diluted. Why not dilute weapons grade nuclear materials to fuel grade nuclear materials? Such a move might also consume some high level nuclear waste. It would also provide energy for an increasingly demanding human race.

Probably the most universally available energy source is geothermal energy. It is everywhere because not far below its surface the earth is reasonably warm- about 60° F at 25 to 100 ft. down. This could supply a substantial amount of the heating needs of many buildings and structures. It would also supply air conditioning for places like Texas where air conditioning uses twice the energy that heat does in Vermont. It could supply from 20% to 80% of the heating needs of many buildings. However, its installation costs are so high that it is not cost effective in individual houses or small buildings, nor is it easily added to existing structures.

That emphasizes a major shift in the energy supply of the future. Everyone should and will have multiple energy resources. The risk of depending upon a single energy source that either fails, like in a hurricane, or is controlled by a single profit-hungry source disappears if you have multiple sources of energy.

There are other types of geothermal energy. Hot springs and thermal geysers are concentrated in a number of special places around the world. These provide opportunities to build geothermal power stations, such as those that make all the electricity in Iceland, and much of it in Norway. We have a few such stations in the U.S. Most are neglected so far. That will have to change, and it will not be necessary to tap Old Faithful for energy.

Solar power is another underutilized resource mostly because it is intermittent, inefficient and expensive. Many of us have a small solar powered hand calculator. Solar powered hot water systems and electric generators exist but are expensive and intermittent. We extract only a small percentage of solar energy with most of our solar devices. Nonetheless, whole homes are powered by solar energy. That is just one more example that shows that the ultimate source of our energy is primarily sunlight. It is plentiful and universally available on an intermittent basis. We have developed some remarkably sophisticated solar energy systems that remain largely unused. The sun provides more energy every day than all mankind uses. We have learned to capture an insignificant amount for practical use. We are still learning, and have much new science that has yet to be converted to technology.

In new construction, just the placement and orientation of the structure can have substantial effect on its energy efficiency. When clearing the land, cut down only the trees on the south of the structure, to allow for passive solar windows on the south. Place fewer windows on the north, and lots of trees. Insulating the foundation saves quite a lot of energy. Sealing air leaks saves vast amounts of heat.

Wind power is another facet of solar power. It is the uneven heating of the earth by sunlight that causes wind. As long as the sun shines, wind power will be available.

So far most wind power has been operated by major power plants. In California there is a huge wind farm of over 600 windmills. Much of

Western Europe, like Norway, Denmark, England and Holland use wind power from windmills out in the ocean. Why not the west coast of the United States?

One problem with our present system is that it puts more electricity onto a grid that is controlled by a small number of commercial ventures. A major objective of energy policy should be to wrest much of that control from monopolistic mega companies and move it to smaller enterprises or individuals. Do you believe that gasoline would be $4/gal, if there were 200 competing oil companies instead of seven? If a substantial portion of our home energy consumption came from individual wind generators, massive blackouts would be far less of a problem. Make no mistake; our electric power companies have been remarkably responsive to public need and to uninterrupted power flow. However, they are motivated largely by the economics of the bottom line. Remember the California energy crunch of 2000/2001? Having substantial supplies off the grid would ease the problems in times of blackouts from major snowstorms, earthquakes, floods or hurricanes.

Wind has its problems. It is unpredictable and extremely variable. In addition our use of wind has been dominated by industry. They want to build huge windmills and make big wind farms to feed energy onto the grid so that they retain control. The big windmills, although efficient when they are operating, won't operate in winds below 8 miles per hour—a condition that occurs less than 30% of the time in most of the U.S. One cure for that is to design them to work at only 2 or 3 mph. That will essentially triple the operating time of a wind generator. Such a windmill will be less productive when operating in low winds, but will operate three times as much time. Large windmills should be at sea where the wind blows stronger, closer to the surface and much more steadily.

There is a hidden reason for this windmill problem. Our society is run on the alternating current (AC) developed by Nicola Tesla—electric current that reverses its flow 60 times every second (50 times in Europe.) The reason we use alternating current is so that we can transmit it long distances. The friction (resistance) of electrical conductors depends upon the flow (amperage) of the electricity, but is independent of the force (voltage) of that electricity. The power (Wattage) transmitted on the other hand, depends upon both the voltage and the amperage. So we can minimize the friction loss (line losses) by greatly increasing the voltage, but keeping the amperage low. Voltage of alternating current can be easily increased or decreased in a transformer. That's why high-tension wires carry 303,000 volts--to transmit lots of power with very little line loss. They can send power from New York to Washington. That could not be done with Thomas Edison's direct current (DC), because a transformer won't work on DC.

Making electricity for the grid not only requires that it be alternating current, but that it is exactly the same frequency of the grid and also exactly in step (phase). It has to be positive at the same time that the grid is positive and negative when the grid is negative. This requirement places serious limitations on windmill operations. These limitations are one reason windmills don't operate below 8 mile per hour.

What if I want to make and use electricity in my own home? I can generate 12 volt DC on my roof day or night and store it in a battery, which I cannot do with AC. I can use it any time I need it. These small windmills could operate at 3 mph, and triple their productive time if you put a cowling around its periphery and capture more wind. (This isn't practical on huge windmills like those on wind farms.) I could charge my hybrid or electric car with one on the garage roof. The growing trend in electricity is low voltage. The low voltage flat screen TV has replaced the high voltage cathode ray tube. Motor vehicles use 12 volts. Lighting is moving toward low voltage (and very expensive) Light Emitting Diodes (LEDs) and the electronics industry is moving to ever lower voltages, so that things can be miniaturized. Computers run on 5 volts, and telephones on 17 volts. Six or twelve volt DC is the wave of the future.

Actually wind power is two different industries. The technology and the equipment for the massive windmills on the mountain that feed the Grid are almost completely different from the one on your roof. The one on your roof is light weight, replaceable and makes DC. The one on the hill is large durable and makes AC. Each serves a separate market. So far the small market doesn't even exist. It will in the next thirty years.

We should create an industry that builds small windmills (400 to 1500 watts) that work on your rooftop. The complaints are that they make too much noise (a soluble problem), or that you will have to rewire your house. Don't rewire your house, because you will still use power from the grid. Just add a few wires for the DC so you can run LED lights, computers, cell phones or small electric motors. You need not go to the grid for power and when a major power outage occurs, you will still be operating. The objective is not to supplant the grid, but to supplement it. That way everyone is a winner. These generators will not supply all your electricity, but about 30% to 70% of it. That's pretty good savings, and it keeps Con Edison in business.

Tides are a little-used source of energy. Near coastlines tide derived power could add to the complex mixture of energy sources. Technology for converting tidal power into a useable resource already exists. In Moncton, New Brunswick on the Petitcodiac River, a dam and causeway do exactly that. Unfortunately, that dam has almost eliminated the great tidal bore that fascinated millions of visitors until the causeway's construction around 1968.

Tides rise and fall twice a day all around the world. Many shorelines have "breakwaters" to minimize erosion and control tides effects on the shoreline. We should install hundreds of small water wheels in these breakwaters, so that the incoming tide will generate electricity as it arrives and sill more electricity as it ebbs. Tides, however, tend to be economical on a community basis, rather than on a home by home basis. Towns and communities should build and operate such systems. There is a company, Voith Hydro, that is installing tidal generators in Spain and in France,

Waterfalls, like Niagara Falls, are other sources of solar power. We have also used dams, like Boulder Dam, but they have other environmental impacts that make them less desirable. We are learning to cope with those problems. Opening the gates periodically, simulating the spring floods, has already made considerable restoration of the downstream environment on the Colorado River.

Energy abounds. We have capitalized on almost none of it, because oil was easier. Oil's heyday has been. It is time we looked seriously at alternatives that have been touted for decades, but remain largely undeveloped because oil was too cheap. The subsidies and tax breaks won by the oil industry should be redirected toward development of alternative energy. When we have an oil crisis, like the political instability of the OPEC nations, of course the oil industry wants permission to drill in The Alaska National Wildlife Refuge. That will increase their dominant control of our energy supply. Take a look at Prudhoe Bay. That is what our wildlife refuge will look like, not just for ten years, but forever. We have left the development of alternative energy sources to commercial enterprises. To increase the bottom line, they have taken the oil route. It is time that the government, which supposedly is more interested in the long term welfare of the nation than today's bottom line, should subsidize alternate sources of energy instead of giving huge tax breaks to the country's most profitable companies. In fact, it might be a good idea to impose a small, but progressive carbon tax to make us think about alternatives to oil and coal. (Sarkozy has done that in France.) Oil's days are numbered and we must wake up to that fact before a crisis arrives-which may not be that far in the future with China and India, coming into that market. If 20% of our energy came from oil, rather than 80%, we would have no crisis today. We have used more oil than we have discovered for over thirty years and that is why the battle for energy gets so intense. It is almost certain that the calculated oil reserves in Arabia and the Mideast are grossly overstated. That means that the oil crisis is more imminent than we want to admit. So why delay finding alternatives? So far it is merely to preserve fat bottom lines for our current energy suppliers—at the expense of the consuming public. It is worthy of note that as the price of a barrel of crude oil increases so do the profits of the oil companies. What is their incentive to keep the price of oil down?

We also have a variety of technologies that can contribute to our worldwide over-demand of energy. Coal can be converted to liquid fuel, including diesel fuel. The Fischer-Tropsch process is over 100 years old, and converts coal to liquid fuels. Hitler used Fischer-Tropsch to fuel his massive assault on Europe, because he had no oil, but plenty of coal. Even older is a producer gas process, which is an inefficient method of making a low-grade (and highly toxic) fuel gas from coal, air and steam. This process may be rejuvenated if we convert to electrolytic hydrogen because for every pound of hydrogen you make by electrolysis you also make eight pounds of oxygen. The oxygen could be used in place of much of the air in the producer gas process to greatly increase the heating (and commercial) value of the producer gas.

To understand some of this, recognize that coal, oil, gas, peat and wood are all finite sources of stored solar energy. Living plants extract CO_2 from air and convert it to cellulose and other vital components of life with solar energy. When those entities die, they are gradually converted by microbes into hydrocarbons like gas, oil, peat and coal. In that process, methane gas is emitted into the air. Methane has by far the largest content of hydrogen (25%), so as they age the hydrogen content of these biomasses declines. Natural gas is largely methane but still has only 20% hydrogen. Gasoline has about 15%, fuel oil has about 13% and coal has about 8%. As the hydrogen content decreases, the viscosity of the hydrocarbon increases until we come, finally, to coal. There are reasons why you want the hydrogen content as high as possible. First, hydrogen provides more than three times as much energy per pound as carbon. Next, and most important, hydrogen is non-polluting---it emits no greenhouse gas like CO_2. (Strictly speaking, that is not true, because water is a greenhouse gas which absorbs heat, but it is already in balance from thousands of square miles of ocean surface. If there is too much it just rains, or snows, producing more streams and waterfalls to tap for energy). Also, hydrogen is a truly unlimited source of energy, since it forms water when it burns. It can be re-created by hydrolysis using entirely renewable energy resources. For that reason hydrogen deserves serious development. The largest barrier to universal use of hydrogen is storage and transportation. Hydrogen has such a low boiling point (-423°F) that it cannot be liquefied economically except in very large volumes. It can be, and is, compressed to a high pressure and shipped in heavy containers.

We must be careful in our search for alternative sources of energy to move toward truly unlimited sources like sun, wind, ocean currents, geothermal or tides. The use of coal, oil sands, natural gas and nuclear fuels are all good alternatives, but they are also finite and will become depleted. Biodiesel is another viable alternative, but recognize that there just isn't enough of it. At the very best we can expect it to supply about 8% of the diesel market. The same is true of corn-derived alcohol. There just isn't enough corn. But worse is the problem that taking a product, like corn, from the food market, and using it in another causes major economic disruption. The price of beef has gone up substantially, and so has bread, as we convert acreage from wheat to corn to capture government subsidies for ethanol. But in addition, biodiesel and alcohol produce more CO_2 per unit of energy than oil. And ethanol is a far worse greenhouse gas emitter than gasoline, because it is made by fermentation, which produces as much CO_2 as it does alcohol even before you put in your gas tank. We should not subsidize fermented ethanol, because there are far better ways of attacking the energy problem.

In summary, remember a few major principles. There will not be a single replacement for oil as our primary source of energy-there will be many. As we move toward new sources, focus primarily on infinite sources such as sunlight, tides, geothermal or wind. Finite resources like coal or oil sands or nuclear will remain valuable resources, but will no longer be our only resources of energy. We should aim to have more than one source for every energy need in our lives. When acquiring new energy resources make sure to calculate the expected payback period. How long will normal usage take to save enough money to pay for any new equipment? The sun supplies more energy to the earth than our current energy demand. To date, we only use a tiny fraction of that abundance. That is very disturbing because the science to capitalize on most of those resources has been known for 100 years or more.

These are a few of hundreds of solutions to small parts of our oil dependence. There are already many more, and the field is ripe for still others. If every one of us in this nation focuses on getting the job done piece by piece, we probably could cut our oil demand by half within five years. Why not? But it will take all of us. In addition, there are other benefits. Creating new sources of energy will also create jobs and employment. For example, we have many new technologies that are in the earliest stages of development. Why can't we take some of those to

China and help them grow their economy not by demanding more oil and coal, but by using our new and developing technologies? That will greatly boost our relationship with the largest nation on earth and will generate much international commerce to help reduce our outrageous international trade deficit.

There is another issue with our alternate energy efforts so far. Most of these are scientific approaches, which are proving that it can be done. The owner advertises that he is an "environmentalist." Most alternate energy systems are elite and high profile today. But we need practical solutions, not fancy ones. We have neglected the most important reason for alternate energy-to keep the costs reasonable. If we look at many of the highly publicized alternate energy schemes, they are scientific wonders that impress people. But they are expensive! We need to evaluate our change to alternate energy on the basis of payback period. How long will it take for the energy savings to pay for the device that achieves it? Those huge windmills have payback periods of 12 or 15 years, which is why they need subsidies and grants to be reasonable. We need low- cost, mass produced alternate energy devices to bring everyone aboard. It is those devices that will create thousands of jobs in the future. We aren't really on track toward that yet. That's what this book will try to reveal.

In making these moves in the energy supply we have failed so far to distinguish between escaping dependence on oil from unstable countries and reducing greenhouse gas production. These are both critical issues, but they are not the same issue. A few moves can relieve both of these problems. But that does not invalidate the use of others for specific purposes.

To reduce global warming, we have focused entirely upon cutting back emissions. We have neglected the other half of the equation. How do we get rid of the greenhouse gasses that already exist? Atmospheric CO_2 has increased from 280 ppm to 390 ppm just in the last 100 years. That is the highest it has been in at least 100 million years. The oceans have absorbed most of our man-made CO_2. A major factor is trees. They absorb CO_2 from the air. The trees of the United States contain more carbon that all the CO_2 in the earth's atmosphere. Why don't we establish a requirement that every school child will plant one tree every year he is in school? That will be 50 million trees a year--about as many as they are cutting down in the Amazon Jungle in Brazil. It is bigger than that. About 50% of our greenhouse gas emissions are absorbed into the ocean. But CO_2 forms an acid in water. We are gradually converting our oceans into a giant pool of soda pop, and we are destroying much marine life in doing so. Oysters and clams have problems making their shells, and corals can't generate their skeletons as easily in an acid ocean. As we take CO_2 out of the air, the CO_2 in the ocean will return to the air where it came from, and make our reduction process much slower, but much more beneficial.

The main point of all this alternate energy is that we all need to do something. It is not enough to let government and big business do it. For example, if we were to introduce gas rationing along with reducing our national oil imports by 1% per month, the alternate energy business would explode. Think of the jobs that just that one move would create. Unemployment would drop by half almost overnight. What would that do for our potential recession if the money we saved on imported oil were spent exclusively on alternate energy? What would that do to the financing of Islamic terrorists in Arabia?

So far, we have not really even formulated a plan for reducing global warming or reducing dependence upon foreign oil. We talk a lot, but do little. Most of us think that it is the job of government and industry to solve our energy problem. It isn't. It is ours. We don't even have a goal! For example, we should set two goals: 1. Pick a year, such as 2012, when we show a decrease in total greenhouse gas emissions. 2. Pick a year, such as 2020 when we see a decrease in global CO_2 from 0.039% to 0.035%. Then we know that we are on our way to the final goal-0.028%. These are tough goals, but they are achievable. It will take all of us!

One final note. Not all of our oil problems come from politics or technology. If we continue to expand world population at 3 to 4% each year, there will soon come a time when no available resource or technology will be able to meet that energy demand, let alone our food demand. Malthus was right! Are we prepared for that?

Chapter 3

Conservation

It is bad enough that we ignore or neglect over 90% of our energy resources, but worse, we squander the ones that we do use. By far the fastest and most effective way to attack our impending energy crisis is conservation. We use about twice as much energy as we need and probably closer to three times as much. That is primarily because of lack of understanding or ignorance. We make little effort to teach our own people about energy conservation. Consequently most of us believe that the energy problem will be solved by government, by energy corporations or by someone else. The most effective and fastest way to do it is to stop the waste. Not Big Oil, not Uncle Sam or Big Auto, but you! That approach is immediate, it costs nothing, it solves both the global warming and the foreign imports problems and can stop the rising cost of energy. You don't have to buy a piece of equipment or replace one you already have. Ultimately you may want to do that, but it is not a requirement.

Conservation involves the massive re-education of our entire population.

Very simple measures like turning down your home thermostat 2 or 3 degrees and wearing a sweater or housecoat will save a surprising amount of energy. Turn the heat or air conditioning off in rooms that are not occupied. Plan your trips to the store or school, so we have one or two less every week. Car-pool with another parent, a coworker or a shopper. Take public transportation, or ride a bicycle in summer. None of these is difficult, but each is remarkably effective in conserving energy.

It works. At the beginning of 2006 gasoline cost about $1.80/gal. By June of 2008, it had climbed over $4.00/gal. And what happened? Americans drove about 10 billion miles less in 2008. We cut back on our gasoline usage so much that the price declined from $4.00 to $2.00 in six months and continued to decline. That decreased demand caused the international price of crude oil to decline from $147/bbl. to $45/bbl. in six months. The speculators, who drove most of that price increase, couldn't make money in a declining market, so they no longer bought crude oil for resale at higher prices. The law of supply and demand works exceptionally well in speculatively manipulated products like energy. If we reverse our action and start to drive more and demand more gasoline, the price will go right back up again. We already have, and it already has. It went up a dollar/gallon in about six months because we started driving more. OPEC cut back production to force the price higher, and will do so again if we give them a chance. What we need is to use less energy! Our answer is to buy less so the price stays low. Alternative sources are a major part of that change.

Household energy demand is the other major consumer of energy—about equal to transportation. In recent decades we have developed some remarkable home energy conservation programs that can cut up to 80% of your home's energy footprint. Many of these can be retrofitted to existing homes and all of them can apply to new construction. Putting insulation under a new home foundation can save quite a lot of your home's energy losses. This is particularly effective in cold climates where heating the home is the largest consumer of energy during cold months. New and old homes can have two inch insulation added around the house. Since heat rises, adding 12 to 24 inches of insulation over the top-story ceilings saves vast amounts of energy. Probably the most cost effective thing you can do is to seal and eliminate essentially all air leaks that carry out much of your expensive heat in cold weather. (They also let in heat in hot weather.)Caulking and taping are time consuming, but add minimal expense. An airtight house has remarkable ability to retain heat energy. Windows are a major path for lost heat. New windows with R factors of five or seven greatly reduce heating costs over the standard window with R factors of two or three. Storm windows add further protection against heat loss.

But be aware of another factor in an airtight house. We need fresh air. If you really have an airtight house the air inside becomes stagnant and polluted, so an air transfer system becomes necessary. These take the heat from the air they expel and put it into the fresh air that they draw in. The best ones also take heat from the moisture in the expelled air.

There are many approaches to conserving energy. But the most important factor is to make a plan. Set a goal of, say, 10% or 20% decrease! Study your present usage and figure out how to modify it to use less. This costs nothing, but does take time and effort. It will save more money than you think, because as we all decrease our demand the price of energy will start to go down! You use energy to heat your house, to drive your car, to heat water and to cook. Make a plan for each of these to cut back on the usage. That plan will enable you to cut your energy expenses without having to scrimp. You will make more efficient use of your energy resources. You might learn a few fascinating things.

Next, check the efficiency of all your present energy consuming equipment. When was your furnace or hot water heater last cleaned and made efficient? Is your insulation adequate? Adding insulation may require a little calculation. If you already have 16 inches of fiberglass insulation in your ceilings, adding 12 more inches may not be cost effective, but if you have only six or eight inches, it will pay for itself the first winter. Heating and cooling your house are very large portions of your total energy usage.

The next level is to modify, or replace your existing system to make it more efficient. Since this costs money, this plan requires a "payback" approach to energy conservation. If you are going to spend money to reduce your energy consumption, calculate how long it will take for the savings to pay back the cost of the change. This critical measurement is missing from almost every energy device on the market today. The windmills you see on your skyline are supposed to decrease our energy costs. Their payback periods are often twelve to fifteen years. (Getting payback figures from the electric utilities is somewhere between painful and impossible. I suspect many don't even know themselves.)Paybacks should be two to five years--the shorter the better. Energy saving devices should be rated in payback years--honest ones, not the inflated numbers you get from public utilities or equipment manufacturers. If we all do it, we will always surpass the plan, because energy costs will decline as we use less. We learned that in gasoline in 2008.

Waste comes in many forms. One of our biggest personal energy users is the automobile. It is also the most wasteful. The American auto industry makes cars with far too much unneeded horsepower. They have resisted making more efficient cars. Starting in the 1940's, they spent billions of dollars improving the efficiency of their engines. Did they make cars that went further on a gallon of fuel? No! They made the engines larger and larger. Then they spent more millions of dollars advertising that unnecessary and rarely used horsepower. Why go from 0 to 60 in four seconds? And where do you do it? Not on my street!

Few cars need more than 75 horsepower, yet even fewer cars have engines as small as 75 HP. A 75 HP engine is capable of moving a car about 45 miles per gallon of gasoline. But automakers are touting their "fuel efficient" cars that get 27 miles per gallon. How do they measure fuel efficiency? It assuredly is not by the distance a car will go on a gallon of fuel! Besides, the numbers they give are false. They assume 100% gasoline fuel, but U.S. gasoline today contains 5% to 10% ethanol, which gets two thirds as much mileage as gasoline. It may be worthy of note that the 22 Horsepower Model T Ford ran on ethanol in 1908 when it was introduced because gasoline was still new and relatively unavailable. Today, because of the agricultural lobby, you are subsidizing ethanol for your gas tank despite the fact that it is a far worse polluter, and much less energetic than gasoline. And look at the side effects. Bread prices went up 75% in less than two years, as we took farm acreage out of wheat and put it into corn for the ethanol subsidy. Beef prices doubled, and corn tripled. Besides, there aren't enough acres to produce all the ethanol needed to drive our nation's gas-guzzlers. That is the wrong approach! Is it any wonder that Japanese make more of our cars today than the Americans? Even they have joined the horsepower race established in Detroit, rather than the economy route proposed by the late George Romney at American Motors, and now more common in Europe.

Transportation takes up far too much of our energy demand-- around half of it. We have built our society around the automobile. In much of this country people literally cannot survive without an automobile. We move further and further from our businesses and shopping areas and become ever more dependent upon the automobile. It will take a lot of planning but relatively little effort for each of us to cut back our driving by ten or twenty percent. That is your job. How often can you save money and contribute to the well being of your society at the same time?

We should institute gasoline rationing. People will have panic attacks at that idea, but if we do it right the limitations can be easy at first and get a little more stringent every year. A 2% decrease the first year, and another 2% each year for another ten or twenty years would work wonders. The best thing about that is that it gets people involved and thinking in terms of conservation. It will be an education process. It should be remarkably effective and surprisingly painless. Would you like to start a Gas-saver Club or an Energy Conservation Committee in your community?

What else can you do? Get a smaller or less powerful car. You buy fuel for a massive engine that doesn't use that power 98% of the time. American cars are designed as (and advertised for) status symbols rather than for transportation. Don't get suckered in. A powerful car doesn't change who you are, but it does use lot of unnecessary energy.

We live in the age of Information Technology. The Internet and the cell phone dominate not only our communication but our lives. Why not make arrangements to work at least one day a week at home and send in your work by E-mail or on the Internet. Even one day a week will cut 20% of our transportation costs, and almost 10% of our entire fuel demand. And think of the traffic jams you will prevent or avoid. That's $70 billion per year that we use here instead of financing King Abdullah, Ahmadinejad, Putin and Chavez. Unless you are on an assembly line, or doing physical labor, almost everyone has some tasks that don't require working directly with your co-workers. Consolidate those hours into one or two days per week at home and save fuel and pollution. Have your staff meetings on line. Even the Polar bears will love you.

There are other tricks you can do. If you inflate your tires another two or three psi, you will noticeably improve your mileage. If you live on a dirt road, the ride may be a little rough, but you can slow down. Plan your driving carefully. When you're going downhill, take your foot off the accelerator. Coast up to red traffic lights and stop signs. Ease up on that gas pedal—if you over use it, you will also overuse your brake, and every time you put your foot on the brake unnecessarily, you waste fuel. Minimize that waste.

But the car isn't the only fuel waster. Do you make a real effort to turn out lights when you leave a room? Or do you use fewer, smaller, or more efficient lighting? Do you arrange your workspace and your work hours so that you use daylight instead of artificial light? How about switching some of your lighting to LED's that are much more efficient? Don't use a clothes drier when the sun is shining. Use the dishwasher less, with fuller loads (or not at all). These things are easy to do, but we don't do them. Make a plan!

Here's a suggestion. Take a look at your last electric bill. It should list the number of Kilowatt-hours you used. Make a resolution: I will cut those KWH by 10% beginning next month. With a target and a time frame things will happen because our focus is on a problem that has been too easy to ignore for decades. We have to wake up.

Heating and cooling your home is your biggest personal energy consumer. (In areas with long commutes to work transportation probably exceeds this.) We have been terribly inefficient in the past about heating and cooling our homes. To begin with, clean your furnace and get its efficiency as high as possible. Simply turning your thermostat down requires no capital outlay but can reduce your heating bill by 10% or more. It is easy to put on a sweater, a house jacket or a robe and keep your temperature three or four degrees lower. If you don't have storm windows, add them. (Remember the payback.) They will save you substantial heating bills. These actions and others of your own choosing will seriously cut your home heating bill and your carbon emissions.

Raising the temperature setting on your air conditioner will have an even greater effect because cooling air costs twice as much as heating it! Do you turn your thermostat down or your air conditioner off at night? Put them on a timer that inactivates them when no one is using the room. These simple things have a remarkable effect on energy consumption. If the efficiency of your home heating system is less than 83%, consider replacing it with a system having efficiency above 90%. (But remember the payback number!)

We already have the tools to cut our energy consumption by 20% without any serious difficulties or major capital expenses. We merely have to train ourselves to capitalize on ideas and technology that have been around for many decades. And if we are good at it we will develop other ideas to reduce our energy consumption further. 10% is easy, 20% is achievable, and 30% should be a long-term goal. Most households spend at least $5000 per year on energy. A 30% energy savings could pay the interest on your mortgage.

There are other things we can do. A prime goal should be to have multiple sources for every type of energy you use. This has two advantages. If a source is disrupted, such as a storm that takes out the electric power, you are prepared to switch to another source. But also, if some supplier, like OPEC or Ahmadinejad, wants to gouge you and charge higher prices you can tell him to go fly a kite while you use another source of energy. That ability alone will help keep energy prices down. We have already seen how OPEC increased the price of oil during which time the oil companies made the largest profits ever recorded in history. Big Oil had no incentive to keep the price down. Had we been able to switch to geothermal, wind, or solar power they would have made a serious effort to keep the price of oil down to keep us buying their product. They won't do it themselves, but we can force them to. Ultimately YOU hold the power of the purse. Give yourself the tools to use it!

There are dozens of other options available to lower our energy consumption. The most important factor in all this is to make up your mind that you take energy seriously, and set goals for specific reductions in both cost and usage. Recognize that in the past thirty years energy has grown from a minor family expense to a major one. Effort today will have a much larger effect on the family budget. It will get bigger. The key is involvement. Con Ed and Big Oil will not do it unless they can make a profit. Government won't do it unless you make them, because Big Oil spends millions of dollars every year lobbying congress (and advertising on TV) to favor them and their approach to energy--the approach that keeps them in control. We must educate ourselves to begin to control the cost of energy instead of letting it control us. A good place to start is in school. Insist that your children get a course in energy conservation in the classroom. They might come home and push you into the right pathway. "Hey, Mom, the sun is shining. Why do you use the clothes drier? " "Hey, Dad, you didn't turn off the lights when you left." A worthy goal?

Conservation is clearly the most effective way to stop our profligate use of energy. That is not the only objective. Our current energy supply is almost all from the burning process, the one that produces the greenhouse gas CO_2. Most of these things will also reduce our CO_2 emissions, but what about the CO_2 that is already there? Since 1900 man has increased the CO_2 content of the world's atmosphere from 280 parts per million (ppm) to 390 ppm. It is still increasing. The problem is even bigger than that. Most of the CO_2 is absorbed by the oceans, which are in balance with atmospheric CO_2. There is about four times as much new CO_2 in the oceans as there is in the air. So cutting back on our emissions will help, but will be painfully slow as the CO_2 in the ocean returns to the air from which it came. The CO_2 in the oceans is already causing problems, as corals and shellfish find it ever harder to form in an ocean made acidic by our CO_2 emissions. That makes it apparent that we must do something about the existing excess CO_2. What?

I was surprised to learn that the biggest emitter of greenhouse gas is forest fires, not the automobile, which runs a close second. They are a far more serious threat, because not only do they emit massive amounts of greenhouse gas, they destroy our most prevalent and successful greenhouse gas consumer--the tree. Trees take CO_2 out of the atmosphere (about 4 billion tons per year) and build the construction materials of life from it. As we burn down our trees, we must replace them. Trees cover about 30% of our earth's land today (somewhere near three trillion trees), but covered over 50% only 200 years ago. I suggest a national school project that requires every public school student to plant a tree every year they are in school as a part of their school curriculum. That's 13 trees per child, and 50 million trees per year. We are cutting 10 billion trees per year world-wide and over 100 million in the U.S. If each student planted 2 trees, we might stabilize our already depleted American forests. This is no small matter. Planting trees in Vermont is one thing, but planting trees in New York City, Chicago or Los Angeles is quite another matter. We will need some careful planning and we will need experts on where to plant what trees and how to do it well, so they will survive and grow. Chicago is already planting a variety of plants and trees on the tops of buildings that have flat roofs. I suggest that The Arbor Day Foundation is a good place to look for that expertise. We have agricultural schools and forestry courses. These could also teach us how to reduce forest fires. The threat of forest fires increases as we warm the planet, so we are currently feeding our own dilemma. It is time we gave serious thought to reversing that trend. How about an idea or two from you? How would you plant trees in a city and stop forest fires in the wild?

Conservation is more an attitude than anything else. We can and must learn to be conservation minded. If we teach it in school at an early age, it will follow us throughout life. I suggest courses in energy conservation be taught beginning in about second or third grade, while the creative mind is active. That will engender a conservative way of thinking about energy in our whole population. The vast majority of Americans haven't any idea of what we are doing to our natural habitat. They don't understand, nor want to believe that we are truly endangering our future. If they really understood that, they would not hesitate to take action. We are all involved. Let's all get informed. Let's all be active.

Chapter 4

Geothermal Energy—An Untapped Resource

We are facing the consequences of our profligate use of fossil fuels. Our supply of oil, coal and natural gas is limited. Since the demand for them increases dramatically it is now becoming clear that we are approaching the limit of those resources. What is not yet clear is whether that will be in 10 years or 150 years. OPEC wants us to believe that it will be 150 years, but reality is probably much closer to 10 years. They have been claiming ever-larger reserves of oil for over a decade, without discovering new oil and shipping sharply increasing volumes. Their reserve figures are probably greatly inflated. The oil industry points out that there is still much oil to be discovered, and that the exhausted oil wells still contain almost as much oil as we have already extracted from them. Even they recognize that oil is a finite resource, and that we don't even know how much oil there is. We have been at it for 150 years. That tells us that there are increasing difficulties in extracting unknown amounts of oil from ever more elusive sources. The costs continue to increase, so other sources of energy will become ever more competitive. We may wish that we had not been so extremely wasteful of this declining resource, but now we must live with it—and pay for it.

The most abundant and the most widespread non-solar energy resource is geothermal energy—the heat energy generated and stored within the earth. It is almost totally neglected today. It shouldn't be. It is available everywhere and is essentially unlimited because the earth will not cool any faster if we make use of the energy it is radiating into outer space on the way by. That amounts to about 128 billion Btu/hr. or the output of 1.5 million home heating furnaces running continuously.

If we dig a hole anywhere in the earth, the deeper we go, the warmer the earth becomes. Toward the center of the earth, temperatures appear to be about 1000° F or more. That's a pretty big pool of energy—far bigger than all our fossil fuels combined. It is everywhere, and neglected almost everywhere except in Iceland and Scandinavia where it is the source of most of their electric power.

Geothermal energy is, like most natural resources, unevenly distributed. Where the earth's mantle is thin we have hot spots with quite high temperatures. These result in hot water pools and geysers such as Old Faithful. These can be, and are used for special services like spas, hot baths, and even power plants. Iceland has an abundance of these, and makes use of them to generate all its electricity. The Romans had some hot baths built around some of them, and we have Warm Springs, Georgia, where President Roosevelt took therapy for his poliomyelitis. But most of the geothermal energy is spread around the globe fairly evenly in much less spectacular ways. If you are on land, you are within about 60 feet of a relatively constant 55° F.

Geothermal energy, like all other sources of energy, has its limitations. You will not produce steam with most geothermal energy because it is not that hot unless you live next to a hot spring. But there are some technical tricks with which you can heat your home. You can also cool your home if it is hot outside. Currently geothermal systems require an air heating system. If you heat with steam or hot water, geothermal energy may require an extensive (and expensive) redesign of your heating system. Here is an opportunity for new ideas and new technology. If you live on granite or hard rock, the expense of drilling down to the heat source may well be too great and your payback period might be excessive. That leaves only about 1 billion eligible homes and other buildings on this earth.

Let's take a look at how a geothermal system works. There is a constant supply of warmth inside the earth. Digging down to that makes available an almost unlimited source of that heat. A heat exchanger and some piping could keep your house from getting colder than 55° F, but that is not a comfortable room temperature. However, 55° is warm enough to boil several refrigerants whose gases can then be piped up to the home. If you compress a gas it gets warmer. You can compress those refrigerant gases until they are about 120° to 150° F. That can easily heat your home. When the refrigerant gives up its heat it condenses to a liquid and is piped right back into the ground to repeat the cycle. You will be heating your house for the cost of compressing refrigerant gases. It is similar to running your refrigerator backwards.

Because of the excavation, the installation costs can be quite high. They can be minimized in the construction of a new home, since excavation is usually involved anyhow, and the additional cost for the geothermal system may not be high. For an existing home, you need to dig a hole down to the warm earth. That could be expensive, so geothermal heat is more cost-effective in large installations and new ones. It may be too expensive for many single-family homes. Remember the payback factor. If your payback is too long you might wonder if it is a good buy. The payback period could be quite long for geothermal heating of small installations. Your supplier or contractor can probably do that calculation for you. A geothermal heating/cooling system could easily be trouble-free for thirty or forty years, so even expensive ones may be worth installing if you have the money.

You will have little or no additional heating expense, although I recommend a small back-up system if your comfort zone is above 70°. One of the big advantages of alternate energy systems is that they can often be combined to gain optimum results. Many systems, including geothermal heating, are best used as partial or supplemental solutions. That versatility gives alternate energy resources a great advantage. We would rarely have to rely on a single source of energy. We are all aware of the problems caused by a power outage, for example, or if your furnace conks out in the middle of February. If we have overlapping systems, such events become nuisances rather than crises.

Geothermal energy will undoubtedly become a substantial part of our total energy supply. It is everywhere, it is clean and it is unlimited. It lends itself to combination with other energy sources, and is essentially trouble free once installed. You won't have to shut the system down in a dozen years because you run out of heat. It may not be suitable for you but it probably is! To find out, look up geothermal heat on the Internet or in your phone book. You will probably find a local supplier. If you don't, you might consider becoming one yourself. The energy is there. So far, geothermal installers are relatively scarce, and the expense is relatively high for individual homes, but both of those factors are changing, and you might be able to contribute to that change yourself. The expense of digging a hole for the installation of the heat exchangers is also quite size dependent. Building a small unit for a single home may be too expensive, but digging a large one to heat a hospital or office building should be much more cost effective.

But there is more to geothermal energy than heating your home or office. There are hundreds of geothermal hotspots scattered all around the earth. Known as hot springs or sometimes as geysers, these are sources of very hot water that can be converted into other energy forms such as electricity. I already noted that Iceland derives all of its electricity from geothermal energy from hot springs. These hot spots are readily adaptable for large community installations, but less readily adaptable for single home heating.

Volcanoes are outlets for geothermal energy. In some rare cases these can be tapped for geothermal energy at relatively low risk. The key to such sources is to keep the investment very low so the payback period is short. In addition, an emergency plan is essential. If you plan to use a volcano, be sure you get the facts on its activity history, so you don't get washed away in a pool of lava.

Chapter 5

Wind in the Eaves

Wind is a major source of energy. Wind is everywhere. It is also extremely fickle and unpredictable. Perhaps that is why it is greatly underutilized. It is remarkable how little we have done on wind power. Much of the literature about wind was written in the late 1970's and early 1980's. Until recently, there was remarkably little effort on trying to find out how best to use wind energy. Windmill formulas are esoteric and arcane. They don't really explain how they are designed to capture wind energy. The math is frightening and contains many factors that are not clearly defined. There is often no specific objective that the design is trying to achieve. No wonder no one has really challenged the conclusions. There is no way to determine if they did the job they were meant to do, or often what that job was.

There has been a resurgence of creative design for wind (and water) turbines in the past decade. Much of it has been in Western Europe and the UK, but it is a worldwide activity. Most of that activity is in large commercial installations. China builds many of the world's wind generators. There has been a tendency in small private installations toward vertical axis wind turbines (VAWT), because they need not shift when the wind changes direction-they work when there is wind from any direction. However they are substantially less efficient than horizontal axis systems. The two dominant types of VAWT are the Savonius, and the Darrieus. Horizontal axis (HAWT) or vertical axis, there should be more emphasis on the neglected small installation.

Jeremy Gorman

The opportunities for small horizontal axis wind generators are drastically different from those giant wind towers you see on the mountains and on the wind-swept Great Plains. Their very size should markedly influence their design. Those giant designs aren't really suited to small rooftop installations close to the ground in crowded areas. So far, however, most commercial rooftop designs are small copies of the large commercial designs. They are not very productive, which explains why they find very little usage. Few of them can pay for themselves in less than eight or ten years, probably longer.

There are several features that can improve the performance of small rooftop generators that are not available to the giant ones used by utilities. The conditions on your roof are markedly different from those on an isolated mountain or in a wind corridor. You can't change the wind conditions at your home, but you can change the design of your windmill to capitalize on the conditions that exist there. You can triple or quadruple the output of a small generator by surrounding it with a conical cowling 40% larger than the windmill. This cowling not only captures twice as much wind to drive the wind foils, it also accelerates that wind so that it will operate at much lower wind speeds. This not only increases the output at any given wind speed, it markedly increases the hours of operation. On your rooftop, winds are light enough to be in the area considered unprofitable for the Utilities. Remember, the Utilities are limited by the need to feed onto the high voltage AC Grid. That does not apply to the homeowner. I know of no commercial example of this simple and inexpensive enhancement for small rooftop windmills. Such a cowling is not practical for the giant windmills. A massive conical cowling 400 feet in diameter would need to be a sturdy durable structure to withstand the frequent violent wind assault found in those wind channels. It would also hinder their ability to swing with the wind as it changes direction. They would also act as a lightning rod in a storm.

Small home windmills are scarce. Most people feel they cannot make them. There aren't many available commercially and none of them has any enhancements like the cowling. Small household windmills so far follow the same design patterns as the giant utility wind generators. They miss many opportunities to increase their output and their payback.

Household generators should be designed to be low cost and replaceable, rather than high cost and durable. They aren't. Perhaps that is why they receive almost no publicity.

Since the energy of wind is a function of the square of its velocity times the mass of the air, doubling the capture area will send eight times as much wind energy through the blades of windmill. Theoretically, a windmill that operates at a minimum of eight mph wind velocity will operate on one mph wind with a cowling that doubles its intake. Practically, that is not so. The non-existent conical cowling should be a critical component of any small household HAWT generator. Households are almost without exception in low wind areas and usually are grouped together so that each house interferes with the wind flow of its neighbor. The essential feature of a home wind generator, unlike the utility designs, is to operate at low wind speeds. It should be a wind catcher, rather than a modified propeller. It functions differently and has a different objective. Generating low voltages greatly facilitates that, so home wind generators should generate 6-, 12- or 24-volt electricity. Such electricity can't be sent long distances, but on a home it doesn't have to be. The future of wind power will be with a totally different system of electricity more like Thomas Edison's than like that of Nicola Tesla. They are different enough that they are essentially two different industries altogether. The light wind industry is in its extreme infancy—one might even say pre-natal. This book wants to change that. With your help it can. The available rewards are gigantic.

The contribution to effectiveness of the cowling cannot be over emphasized. The low cost cowling allows the small generator to operate in an urban environment. It capitalizes on the very light winds that the large utility windmills ignore. Everything about the cowling is favorable to increasing output. It is a wonder that cowlings have not been in use all over the globe.

An example will illustrate this remarkable benefit. A windmill capturing 1 square meter of air (50 inches in diameter) will receive a remarkable 271 watts of wind power at a wind speed of 8 mph. Note that I said it would <u>receive,</u> not <u>produce</u> 271 watts. Commercial windmills of that size are rated at about 500 watts, but they are rated at their optimum

Jeremy Gorman

speed of 28 mph. At 28 mph that windmill receives 3320 watts of wind power. It converts only 15% of its wind power into electric power even at its rare optimum operating conditions.

The design of the windmill will greatly affect the actual output we can extract from that wind. Current commercial designs convert only about 3% of the wind energy to electricity. If you made the windmill 1.4 times as big (70 inches in diameter), you would double its throughput to 542 watts at 8 mph, and 6640 watts at 28 mph. But with a cowling that is 1.4 times as big as the windmill, you will not only double the air throughput, but double the velocity as well. It will still receive 542 watts of energy at 8 mph, but the windmill will operate at 2 mph. This is far better and far cheaper than doubling the size of the windmill. It is clear that the benefit from the cowling is more from the increased velocity than from the increased mass of air processed. This is even more desirable because it greatly increases the time of operation.

Commercial designs tend to ignore the low velocities because the output is too low to add to the 60-cycle grid voltage. But the wind blows three times as much time at 4 mph as it does at 8 mph. A guiding cone could make a small windmill a cheaper alternative than the public utility. It should be a practical source of electric power for your home. With the use of low, direct current voltage and batteries or other storage facilities, power can be generated almost any time and stored for usage when the demand occurs. Alternating Current (AC) cannot be stored and must be available when the demand occurs, even if the wind is not blowing. Small wind generators are not that complex. Consumer models exist, but they foolishly follow the design of the massive ones that utilities use on mountaintops. They are too expensive at $800 to $2800 (plus installation costs), and take many years to pay for themselves because they don't produce any power the vast majority of the time. The big surprise is how little of the wind energy actually shows up as electricity. I made a calculation of the electrical output of one commercial 400-watt home wind generator that is on the market for $600 (plus shipping and installation costs). I was amazed to see that it converted only 3.3% of that wind energy into electricity. No wonder wind is not relied upon for more of our energy. The manufacturer claims 51% efficiency. 51% of

49

what? I regret to say it but I believe that the current wind generator is built more for show than for performance.

But let's not get carried away. There is no way that a cowling that captures twice as much wind will produce eight times as much energy. Will the wind pile up in front of the windmill and go around it? There is much to be explored in the design of the cowling, and the windmill blades. A very steep angle in the conical cowling will probably act as a barrier to the wind and make it go around the outside. But a very shallow angle will make a very long and cumbersome cowling that may be subject to wind damage. How much of the air can the propeller blades capture before they too will block the flow of the air? Here is an opportunity for Joe Handyman to redesign and test configurations that will maximize the output. A worthy effort could well put a windmill on every home in America and in much of the rest of the world.

I have designed a wind generator that should convert 15% or more of the wind to electricity even at its lowest operating range. It should cost around $400.

There is another facet to this advantage. Designing a turbine or windmill to work on low wind velocities allows you to put your device close to the ground even in heavily populated areas. A smaller tower with guy wires to stabilize it is perfectly suited to urban areas. You can plan the placement of your turbine to capitalize on small wind channels created by surrounding buildings and trees. What is generally a disadvantage can be turned into a local advantage if you plan well. This contributes to general appeal of small wind turbines and puts them within reach of city and suburban dwellers for whom current designs are between impractical and impossible. With a light wind generator you need not have a forty-foot tower 300 feet from your home. Such a tower may cost twice as much as the turbine itself. A two to ten-foot one on your roof can take advantage of the small wind channel formed by the peak of your roof or the top edge of a small commercial building.

Our current wind power is a perfect example of a good idea that got lost in a commercial world. Essentially all of the wind generators in existence today are massive propellers on large towers putting electricity into the grid. Even the few small ones follow that same design—a three bladed windmill with narrow blades which are even narrower at the tips, allowing the bulk of the wind to pass right through the airspace without ever touching a blade. They are expensive machines owned by power companies. They are built with government subsidies and grants on fought over public lands often requiring legislative action. Many of them today are built in China. Power companies make money on them by selling the electricity they produce. Since they won't operate at winds less than 8 mph, they are idle more than they are productive.

Our nation's approach to wind power has been completely backwards. We have neglected the best features of wind, and capitalized on the worst features. Wind is everywhere, but we locate giant wind generators in small specific high wind locations. Wind is gentle most of the time but we only utilize it when it is strong. We put up huge windmills that generate alternating current that can be sent miles away to areas that have wind of their own. They are frightfully expensive, requiring 15 to 30 years to earn their keep with the power they produce. The power companies, however, cite payback periods based on optimum operating conditions that actually occur less than 10% of the time and fail to mention the subsidies, grants and tax relief that financed their construction. Their designs are seriously hampered by the fact that they feed onto the Grid that requires high voltage 60-cycle alternating current that must be exactly phased to the grid it supplies.

Electrical equipment is moving to ever-lower voltages in electronics, transportation and in LED lighting. We subsidize huge profit oriented industries that charge you for electricity that you can make yourself for only the one-time cost of the installation of a wind generator. Commercial windmills operate only 30% of the time. Yours should operate more than 75% of the time. Don't feed the output onto the Grid where a power failure cripples all of us. Create an alternative or secondary system that will operate when the grid is down. Matching the output to the Grid requires intricate and expensive equipment and substantial power losses. Don't follow that pattern. It doesn't apply to a home wind generator.

Why are the airfoils (impellers) narrower at the tips where the most wind is? The official explanation is that the airfoil blades create turbulence in the flowing air, which makes the blades less effective. But unlike an airplane propeller that operates at high speeds where turbulent flow is constant, a wind generator often operates at low speeds that create less turbulence. The net result of these design flaws is that most of the wind remains unused even in a working windmill.

The focus on ever more massive structures that require fought-over land purchases and invite NIMBY (Not In My Back Yard) complaints from surrounding citizens is fine for utilities that want to keep control of electricity. Those expensive problems often impose delays of from three to six years. The land purchase and the lawsuits often cost more than the windmills, which are far from cheap. And the time to recover the costs of these massive showpieces is often fifteen or more years, although the utilities tell you it is only about six years. That helps explain the need for government grants and subsidies. The calculations about how long it will take to recover the cost of installation usually are grossly overoptimistic and are based on the costs AFTER the government subsidies, grants and tax incentives. What if one of those mountain top units is struck by lightning and is incapacitated? That happened twice in Vermont--one called Grandpas Knob (the first commercial wind generator in the U.S.) back in 1944, and one in Searsburg in 2008? Since most of them are on high hills or mountains where the wind is usually stronger, the lightning threat is a serious consideration.

Why are large commercial wind generators designed like airplane propellers? There are several important differences between a wind generator and an airplane propeller:The wind drives the airfoil, instead of the airfoil driving the wind. The speed of an airplane propeller is many times greater than the speed of a wind generator. The propeller is designed for a constant (optimum) rotational velocity, but wind is extremely variable. The power in an airplane propeller is one or two orders of magnitude greater than a wind generator. The moving airplane propeller attacks still or stagnant air, but the wind generator cannot reduce the wind to zero because still air can't get out of the propeller. Small wind generator designs should capitalize on those differences. There are newer and more creative designs that do, but they have not yet found a large commercial market. It is worthy of note that even the airplane has largely abandoned that propeller in favor of turbines and jets. The electric utilities might want to consider those facts.

These problems demonstrate that creative new ideas can get led astray and become obsolete and ineffective. Wind power is excellent, but it, like most basic concepts, needs to be thought out and made practical as we learn new things. The concept has been dominated by large corporations. They use lobbying to gain and keep control of energy so they can make money on it. Massive wind farms support that control. But energy is used everywhere and should not be relegated to a few massive corporations. In addition, wind is everywhere, and its output need not be sent miles away to other locations.

All this leads to some practical conclusions: Wind generators should be in one of two places: Over water (commercial) or on the roof of the point of use. These will be vastly different facilities. Homeowner windmills (wind catchers?) should be small, light, inexpensive and replaceable. The large commercial ones should be over water to capture stronger and steadier winds and because no land purchases are necessary. Over water, the towers need not be so tall or structurally demanding. We can minimize, but not eliminate government interference, or NIMBY complaints, because any public venture will eventually end up as a political debate. But the wind generator on your roof will not!

Putting windmills on your roof does involve some accommodation. They should feed a separate set of wires to handle the low voltage DC. This need not be a major barrier, or a high expense. They should not be durable luxury fixtures whose cost will take dozens of years to recover from energy savings, but small inexpensive and replaceable devices that can be operated and maintained by a homeowner. They should be an alternate system, so that you are not without power because of a major storm that is 50 or 100 miles away. They should be mass produced, or even made by the homeowner himself. They should be as common and as cheap as the kitchen sink. It will take thousands of American workers, whose former jobs have been sent overseas, to make these for a mass market. Hundreds of small companies should be encouraged and competition invited to make them ever more productive. That system will also create a large parts and repair market that can employ still more people. If the government is going to encourage alternate energy with grants and subsidies, they should not give money to big corporations that can afford to do their own development of profitable new products, but to the creative American worker who wants to cut his electric bill in half and has a new way of doing it. Help him manufacture and sell that idea. Perhaps our Department of Energy should take a page from Mohammed Yunus' Grameen Bank idea and help bring energy to every citizen with mini grants and small loans. I'm sure government will find it easier to offer fifty massive subsidies to big corporations, rather than fifty thousand small subsidies to creative entrepreneurs. Easier is rarely better.

Of course we will falter and make mistakes as we enter this untested field, but we will learn quickly. That is where you come in. We are in desperate need for local ingenuity. It is a fertile field for new ideas. You can go onto the web and find designs and construction instructions and even parts and equipment. You can buy small wind generators on E-bay, or on Craig's List. You will meet some remarkable people, and can learn a lot and save even more. But the objective is clear. In 20 years, at least 50% of all homes should generate some portion of their own energy demand. Light wind generators should become a substantial portion of that market.

Not all buildings are practical sites for windmills, but millions are. Every eligible home should have one (or more) roof top windmill for the house and perhaps one on the garage to charge the plug-in hybrid, or electric automobile. They should be designed to operate at 2 or 3 mph winds instead of 8 mph, thus tripling the time of operation. If they are less efficient, that's OK, because there will be no transmission line energy losses. If we generate DC, or rectify the AC to DC, we can store it in batteries or capacitors for use when the wind is not blowing. Wind energy on your roof will not replace the grid, but it can reduce your use of it by from 20% to 80%. But even better, it provides you an alternative source of power. I cannot overemphasize that fact that every energy source in your life should have alternatives. That by itself will diminish the influence of power monopolies. If there are no single source suppliers, there will be less price fixing and profiteering.

We haven't even approached all the creativity of wind generation yet. The current horizontal axis windmill is obsolete. It needs improvement. Consider vertical axis generators for your home. In 1922, Sigurd Savonius designed a simple vertical axis wind machine. It consists of a 55-gallon drum or five gallon can cut in half down its vertical centerline. The two halves are moved off-center about 80% of their diameter and mounted onto a circular base twice as large as the drum diameter. When the wind blows from any direction the two halves act as wind scoops and the wind causes them to rotate. One advantage is that they will catch even light winds. We should spend as much time perfecting the Savonius wind generator as we have designing the massive windmills that feed their output onto the Grid. You can stack two or three on the same base. Consider making the supporting base and crowning top larger and conical instead of flat, so they capture more wind to drive a bigger generator or alternator both faster and more often. These units can drive small electric generators by the thousands. Output could also be used to generate hydrogen for use in your furnace, hot water heater, or stove. The ideal situation is one in which a hydrogen generator is a back-up system for the battery. When the battery is fully charged a hydrogen generator kicks in so that as much wind energy as possible is utilized. In addition, since batteries are expensive and heavy, we might consider massive capacitors to store electricity in place of or in addition to batteries. Generation and use of low voltage DC is a fertile field for new ideas. What are yours?

There are other vertical axis wind generators. The Darrieus wind turbine consists of several vertical strips of metal mounted on a spider-like rack, which catch the wind, and turn no matter which way the wind blows. There are dozens of variations of these two vertical axis windmills, many of which can be built for around $200 in your own house. A good part of the design effort appears to have been to get them to work at low wind speeds. I am not the only person who recognizes that we are neglecting far more of our wind power than we use even in our most common designs. Such plans are also available on the Internet. There are others. In England there is a design called Aero Cam that catches the wind either in a vertical or a horizontal mounting and will operate in low wind situations. It consists of from four to eight flat blades hinged on a rotating rack so that they catch the wind as it blows, but align with the wind on the reverse path into the wind. Because of the problem with the return path of the impeller blades, they are less efficient than standard horizontal axis windmills, but they operate at low wind velocities. Hopefully, these and many more will become readily available on the open market.

Recently there have been many developments in wind turbines. Creative designs of both VAWT and HAWT devices abound. On the web you can find at least a dozen, every one of which has some particular capability to make it worth considering. Some of these can be mounted either vertically or horizontally. Most of these are expensive because they are individually built instead of mass produced. The whole idea is to get enough demand that several different designs become small businesses making low-cost light wind generators. The key to most of these is getting them to work at very low wind velocities. If they are made of replaceable instead of expensive durable parts, they will become competitive and their prices will come down. Wind is pretty fickle stuff and making massive structures that can withstand major changes in wind may not be as practical as making low-cost replaceable units in case of damage. Remember, these are mounted on your roof and can become attractive to lightning strikes. If they are made of replaceable plastic instead of durable steel, they will not be so vulnerable. Your interest is the driving force behind such progress. If we do it right, there will be a wind turbine on almost every new house that is built in the 21st century.

Electric utilities oppose this idea and may lobby against it, but you will benefit and should be loud and active in support of it. Think of the jobs it will create. In fact, we could very well build an export industry, and reduce our miserable imbalance of trade. The Chinese are building one new coal powered electric energy plant every week. They are also building huge wind farms like those in California. But China, too, has caught the Massive wind generator bug. I think they caught it from us. They don't seem to have small windmills on homes either. There is a massive opportunity here. Society can live on the light wind that utilities deem too small to use. Perhaps we could teach the Chinese the advantages of small wind power and sell them millions of small wind generators made in the United States. Since they are also building wind generators, they are aware of wind power. Perhaps we can help them get into the small homeowner's wind generator business, instead of the massive public utility facility. Let's not only replace the coal-fired power plant, but minimize the need for it.

In 2008, almost one percent of America's electric power came from wind---a tenfold increase in a decade. Opening up the local small windmill market could easily increase that another tenfold in the next five years. However, like all energy things, it will take all of us. And it will not only pay for itself, it will be fun. The challenge is not "GE can do it", or "Government can do it", or "You can do it". WE can do it! Get involved. Work with your friends and neighbors, because together we can accomplish things impossible for one person alone. Form groups and organizations to capitalize on the diverse capabilities of everyday citizens. Form a local Club 280--that aims to get the atmospheric CO_2 down to the 280-ppm that prevailed for a million years before the industrial revolution. Form a Wind Energy Club, which sets out to make half of its member's electricity from local wind or water mills. Neighborhoods should work together to define and satisfy their energy needs. These would be excellent projects for teenagers who have spare time. We have become too complacent with our aging system of electric power. We need fewer Enrons and more Grameen Banks. The "Mainstreet Power Group" should become a fixture in every town. Whatever you do, do it now!

Chapter 6

Solar Power

Man's insatiable demand for more and more energy will write a huge chapter in his future. Americans are in the Mideast primarily because of oil-the most prevalent current source of energy in the world. We have fought over oil, and undoubtedly will fight over oil again. It need not be so. Let's look at energy from its most basic components. There are four sources of energy on earth today: solar, geothermal, nuclear and lunar. The overwhelming source is the Sun, which throws more energy on this earth every day than we use. It comes to us in a wide variety of forms. Nuclear energy is the next largest source. It comes from billion-year-old atoms that are unstable and release energy as they disintegrate. Its use as an energy source is entirely an invention of man. Next is geothermal energy, which is widely dispersed all over the earth. Geothermal energy, which is the energy generated deep within the earth most probably by disintegration of radioactive elements, is not even 1% of that supplied by the Sun. Finally there is tidal or lunar energy, which is universally available at any coastline, but almost unused. There is a special category of tidal energy found in ocean currents. These derive at least some of their energy from the rotation of the earth in a mechanism called the Coriolus Effect. Except at the equator, moving north or south on this round earth automatically moves you closer to or farther from the earth's axis and causes east or west drift because your rotational velocity is faster, or slower than the earth under you. That is the Coriolus effect.

Many of you may ask "What about oil and coal and wood, hydroelectric or wind power? "These are our major sources of energy today, but each of them is in reality solar energy that has been captured, often by living things, and stored. I mentioned that solar energy comes in many different forms. Fossil fuels are only some of these forms. Wind and hydroelectric power are solar energy in the process of redistribution. The sun evaporates the water, which gathers in clouds in the colder upper air, and comes down as rain that creates waterfalls, streams and rivers. We are utilizing these resources at an ever-increasing pace, and our costs reflect it. But, like the wind, we neglect and misuse much of it. Back in the early days of oil exploration, between 1879 and 1905 there was so much oil, and so little demand, that it was cheaper than water, which was $10/barrel in those dry climes where the oil was most abundant. The waste was colossal. Because of our profligate use of this resource the cost rose to over $140 per barrel in 2008.

Fossil fuels are finite sources of solar energy that will eventually be used up. There is enormous debate about how long these will last, but there is no debate about the fact that eventually they will become exhausted. It took millions of years to generate those fossil fuels, but we use them up in minutes. That will force us to seek other sources of solar energy—our most abundant energy resource.

Sunlight, on the other hand, is unlimited for all practical purposes. It is so widespread and bountiful that we have neglected it for most of history. We let plants and other living creatures capture that solar energy for us and then capitalized on their results. We have by-passed those intermediates very little in our history. We used water wheels, one form of solar energy, long before the birth of Christ, and windmills were a major energy resource of the early American farm. But those were limited energy niches that never supplied the majority of our massive energy consumption. Windmills in Holland have been a major source of energy for over two hundred years.

We are growing wood every day, so it is theoretically unlimited. A living plant is nature's way to capture and store the energy of the Sun. The problem is that we must kill the living thing to recapture its stored energy. We cut down more forests than we grow, and have for a millennium. Cutting down all the forests is far more imminent than the Sun going out. Forests, too, will be a limited resource. Man has already cut down over half of the world's forests.

Earlier I mentioned that the sun sheds more energy on the earth than we use every day. That point cannot be over emphasized. We have an almost unlimited source of energy that is greater than our demand. What's more, it is everywhere. Why are we fighting over energy? The prime reason is that we have not capitalized on the Sun as a true energy source. It is too prevalent to monopolize, so commercial interests cannot dominate or control that resource. Corporations have commandeered fossil fuels not only because they are easy, but also because they are controllable. Chip out a chunk of coal and set it on fire. It is not quite that simple to use oil, which must be dug from deep within the earth to make it accessible. That explains why coal was used for over 800 years before we began to use oil. But advanced technology has easily overcome that barrier, and today, 150 years after it was first drilled, oil is the largest single source of our energy in the world. As with coal, it requires large well-financed corporations to extract the oil, thus placing the major source of our energy in the hands of a few well-organized profit-making organizations. They jealously guard their monopoly and fight to protect it. They advertise its advantages, and obscure its problems.

That is not the only problem with this over simplistic solution to our energy demands. Air pollution, political haggling and devastating oil spills besmirch the oil industry and plague its customers. Further, oil is the primary source of many useful chemicals, including plastics. If we burn it all up, how would we make synthetic fabrics, telephones, plastic bags and computers? There must be a better way.

There are many! One stands out as a primary energy source--- capturing the abundant energy of the Sun directly, just as plants do. By-pass all that plant intervention and the millennia needed to decay them and their herbivorous consumers into useable energy stores like coal and oil. 1000 years ago, when coal began to be a major source of energy, we didn't have either the science or the technology to convert sunlight directly into useable energy. Today, we have the science and many such technologies. These are the tools by which we can escape the political intrigue, the air pollution and global warming caused by oil. But they remain under-utilized. It is time we stepped out of the old regimen, and set a new course in energy supply which will be adequate, non-polluting and non-political. Now this last, non-political, may be a pipedream, but it surely is a worthwhile goal. The fascinating feature of any alternate energy plan is that even the oil company executives know that eventually that is what we will have to do. What are we waiting for? Why do we suffer through air pollution, water pollution, political backbiting and capricious energy costs when there is already present an abundant supply of energy that we have the ability to use? The answer, of course, is because it is easier. Man has a habit of doing what is easier as opposed to what is better. The science of using solar radiation for energy is relatively new--about 200 years old. But the technology to bring that science into practical usage is even newer—mostly less than 100 years old. We haven't put it to use largely because the oil executives have convinced us that they have enough oil to last past our lifetimes and those of our grandchildren. Besides, as oil becomes ever harder to find and transport, the price and their profits will go up. As we put alternatives into use, their profits go down. As of today, they are still in control.

 Solar energy is everywhere, so we can, with our new technology, use it at home or locally. You don't need to go to Con Ed, or Exxon Mobil. You don't need to use oil from Saudi Arabia. Solar radiation arrives everywhere in a very broad spectrum of colors or wavelengths. The very long wavelengths include infrared light that you cannot see, but your body feels as heat. Most of us have used an infra-red light occasionally and each of us has felt the infrared of the sun's rays. There are the various colors of light that we use to see things with our eyes. As the wavelengths keep getting shorter, the amount of energy they carry increases, but their visibility goes down as we come to Ultraviolet light that we cannot see, but which can make special receptive materials glow. As the wavelengths keep getting shorter, we come to gamma rays and other very high-energy radiations that we cannot see, but which can be utilized by specialized instruments, or can damage us with radiation burns. These are all present in sunlight in varying amounts. As the energy content goes up, the percent of solar radiation it carries decreases sharply. We do have a new technology that capitalizes on all of these wavelengths, and we have many devices for each one of them, or each group of them.

In general, we use the long wavelengths to heat things. Solar panels on your roof can heat water that is used to heat your morning shower, or to be piped though a baseboard heating system to warm up the house. We can use the violet and ultraviolet radiation of the sun to run photovoltaic panels that convert a portion of the sunlight into electricity, currently about 20% or less. Scientists are working on methods to increase the percent of sunlight that can be converted to electricity or, even better, to do all these at once. They have not yet achieved that goal. I would like to put a photovoltaic panel on top of an infrared panel, so the upper layer converts the short wavelengths into electricity while passing the infrared to the second layer to heat water or the home. I know of no such system in use today.

We have all seen photovoltaic solar panels. These are devices that absorb portions of the sunlight and convert it to electricity. Many of us have hand-held calculators that recharge their batteries with tiny photovoltaic panels. They are made of silicon, the second most common element on earth. (It comprises 25% of the earth.) Despite that, photovoltaic panels are expensive--around $5 per square foot. A square foot of photovoltaic panel will produce about one watt of electricity when it is in full sunlight. They operate about ten hours per day, so each square foot produces about 10 watt hours per day. Since your house uses about 50 kilowatt-hours a day, you will need about 5000 square feet of solar panel to satisfy all your electric needs. That costs $25,000 plus installation. It also takes up more than twice the area of your roof.

At least part of the reason photovoltaic panels are so expensive is that they are not mass-produced. If they were mass-produced and marketed, the cost should be reduced by about 60%, and it would provide employment for many people. That's still $10,000 for a house. The advantage of such solar energy systems is that they are universally available. But the disadvantages make replacing the "Grid" impractical. Supplementing it should be the goal. Look at the huge power outages we have suffered in the last two decades. That is because we are all on

"The Grid. "We all go to a few resources for our electricity. When that fails, we all suffer, because we have no alternative. Wouldn't it be great to have other electric sources off the grid, and not be dependent upon profitable high-tension wires and power plants for our electricity from distant sources? Yes, but the owners of the grid oppose that concept. That is their source of income, and they don't want it to go away. Getting off the Grid is not the objective. Supplementing it is. We are always better off with alternatives. Putting up photovoltaic panels might involve some fairly substantial initial investment. But maintenance costs are minimal, and far less than we now pay for electricity.

There is another way to make photovoltaic electricity more competitive. The current silicon photovoltaic panels only use about 20% of the sunlight to make electricity. Scientists are working in new coatings that convert a much broader spectrum of sunlight into electricity, as well as lowering both the mass and the cost of the panels. When we are able to utilize this technology in solar panels, we could perhaps double the wattage available from a given panel while reducing its cost and its weight. If we had photovoltaic panels that could produce 2 watts per square foot, we could cut the cost to $5,000 for a household or even less. That is getting more reasonable and more competitive with the increasing cost of electricity. The average home uses about 18,000 KWH per year costing in the range of $2000. Cutting out 2/3 of that with solar panels would put the payback period under four years. Now we're getting more nearly competitive. But, recognize, that we still won't be and shouldn't be completely free of the Grid. Solar panels produce electricity most when it is needed least, and vice versa. Lights are needed at night when the sun isn't shining. When nights are longer, days are shorter, but demand is greater. What if a tree falls on your solar panels and shuts them down? So don't cut down your utility pole yet. Recognize that generating your energy locally requires an entirely new approach to energy.

Solar systems are not without their problems. Among others, they don't produce alternating current. The overwhelming majority of our appliances are geared for AC. The reason is that electricity is made at large power plants scattered all over the country. Your electricity is piped into your home from considerable distances away over the Grid. The Grid was established during "Rural Electrification" in the 1920's and the 1930's. It has succeeded in its objective of supplying electricity to essentially every home in the United States (and Canada). This vast system requires that we use AC, as explained earlier in this book.

But the age of electronics is doing us a favor. In order to miniaturize computers, cell phones and iPods, we are moving to ever-lower voltages, so that very thin insulators can be used in those very small devices. Higher voltages would arc right through the thin insulators. Notice that the low voltage flat screen in both television and computers has supplanted the high voltage cathode ray tube. Cell phones, iPods, digital cameras and Blackberries all operate on low voltages. So does your automobile. The need for high voltage is declining for everything but long distance transmission lines. So getting off the high voltage grid is becoming ever easier and more popular, to the concern of Consolidated Edison. So the door to making your own electricity is opening. With solar panels on your roof, you don't need to send its output far and you don't need either high voltage or AC.

Consider this. One reason for having alternate energy is to have a second source in case of power failure. Your solar panels can supply from 20% to 80% of your electric power, and you will always have two sources of electric power. You can use low voltage DC in your computers, your TV, your automobile, your cell phone and your GPS. There are also 12 and 24-volt appliances already on the market for boating enthusiasts, campers and mobile homes. LED (Light Emitting Diodes) lighting is much more efficient than either incandescent or fluorescent lighting, and is finally coming on the commercial market. They are still exorbitantly priced, but gradually coming down. With a little research and some personal creativity, you can probably save at least half of your electric bill, and cut your greenhouse gas emissions by half. That will require more planning than money.

There are other ways to use solar energy. As I said earlier, photovoltaic panels utilize only about 20% of the energy of sunlight. Although we are developing special Photovoltaic panels that use 30% or 40% of the sun's rays, there are still large portions that won't convert directly to electricity. Engineers have developed and deployed massive parabola shaped solar panels that focus *all* the solar energy onto pipes containing oils that get heated to around 700 degrees. The hot oil can produce steam to power engines or drive generators that produce electric or other power. This technology may outpace both photovoltaic and solar heating panels. They are quite efficient and utilize a much larger portion of the solar energy. They are complex and expensive, so they do not lend themselves to small homeowner installations-yet. They usually are run by large utilities or atop large commercial buildings. Nonetheless, they are a source of low cost solar energy even if they work best for large profit-driven utilities. They produce no greenhouse gases. They produce no income for Saudi Arabia and Iran.

Solar energy is an invitation to more creative thinking on how to make use of it. What are your ideas on getting the sun to work directly for all of us? Contact your nearest university or energy research center. They need new ideas. We all need new energy resources.

Chapter 7

Nuclear Energy

Nuclear energy provides a significant portion of today's electric energy. There are 105 nuclear energy plants in the U.S. and 400 worldwide. In the U.S. they supply about 8% of our electric power. Like fossil fuels, nuclear energy is a finite resource and will eventually be used up. But there are some significant differences about nuclear energy. People are afraid of nuclear energy because of Three Mile Island and the meltdown of the nuclear power plant in Chernobyl, Russia. Despite that the safety record of the nuclear industry, even with Chernobyl, is far better than that of coal or oil, which kills about 500 workers worldwide year after year. The American nuclear energy industry has not killed a single person in the last forty years.

The problem at Chernobyl, like Three Mile Island, was that they were so convinced that they had everything in control that they grew lax and lost that control. What's worse, when they lost it, they weren't prepared with an emergency plan to deal with the consequences. Man has a history of assuming that he can make infallible plans, so he doesn't bother to establish a plan for a failure. That is what happened to the grounded oil tanker Exxon Valdez in Alaska, at Three Mile Island, Pennsylvania and at Chernobyl, Russia. It was the primary problem in BP's disastrous oil spill in the Gulf of Mexico. The ability to deal with the failure existed, but BP had spent 12 million dollars lobbying our government to prevent being required by law to install them.

Nuclear Power plants use "fuel grade" nuclear materials that contain between 82% and 93% "fissionable" materials--materials whose atoms break apart under certain circumstances and release huge amounts of energy. Uranium and plutonium are the prime nuclear materials. Fuel grade nuclear materials cannot undergo a nuclear explosion, but they can get seriously out of control and produce vast amounts of unmanageable energy and radiation as they did at Chernobyl.

The nuclear arsenal consists of "weapons grade" nuclear materials that contain more than 93% fissionable materials. Weapons grade nuclear materials can create a nuclear explosion like those at Hiroshima and Nagasaki in 1945. (See the cover of this book.)Refined weapons grade material is scattered all around the globe in the 11 nations that have them. It is carefully guarded and has severely restricted access to very specialized experts.

Most of us are frightened by nuclear power. We want to avoid it altogether. Nonetheless, nuclear materials abound. Since 1945 the world has mined and concentrated thousands of tons of fissionable nuclear materials. The overwhelming majority of that is in weapons--nuclear warheads. What few people realize is that the nuclear arsenal contains more than 100 times as much fissionable material as all the fuel grade materials in existence. That is an enormous pool of energy whose official purpose is never to be used. The world neither wants nor needs any more nuclear explosions.

These weapons put the whole world at risk, because some despot may acquire some of them, or gain access to them. This is a political issue that has plagued the world for over forty years and has given rise to many treaties that don't really solve the problem. They are often broken.

There is another unmentioned risk of weapons grade materials getting into the hands of despots- Storage! Weapons grade materials will explode if they are in packages of more than a "critical" mass. If you put more than that critical mass in one package, it will explode without any outside stimulus. Allowing such a risky material to get into the hands of inexperienced and untrained people could be a true disaster.

Nuclear power plants consume those dangerous materials. We have the capability of "reprocessing", or diluting weapons grade nuclear materials into fuel grade for our nuclear power industry. We can and should utilize that enormous supply of energy that we created at enormous expense. It would become a major supplier of nuclear fuel. It far exceeds our current capacity for mining more uranium. We could build more nuclear power plants to further reduce our dependence on fossil fuels while consuming large numbers of our nuclear weapons. France already makes 26% of its electricity from nuclear reactors. It would reduce pollution too. There already is a tiny amount of reprocessing of our weapons grade nuclear materials into fuel grade materials. What's so bad about that? Why don't we stop the mining and refining of more nuclear materials so that the weapons will be used up by the power plants? No such plan exists.

This dilemma is an opportunity. Nuclear waste materials and weapons are each a dilemma. But they are made up of materials that have commercial value. Instead of agonizing over how to get rid of these two problems, let's combine them into a commercial enterprise that creates value and earns money. We have ideas on how to make use of the low level waste materials in medicine and instrumentation, but we don't. We can determine how to combine the weapons grade materials with spent fuel rods to make commercial fuel rods for future use. We would be solving two problems at once and creating a new source of energy. Hire our unemployed in a commercial industry that produces nuclear fuel, radioactive materials for medical and instrumental controls and makes some money doing it. Of course we are stalled on nuclear waste, because it produces nothing but expenses—no valuable products. Combining those two problems can produce a great deal of commercial value, and employ our people in doing it. We don't have a clear idea of how to do that yet, but we can learn. We had to learn how to create all that waste. Surely we can learn how to extract the value it contains. Here is another problem that is an opportunity.

Except for the depletion of nuclear weapons, nuclear power currently has no real status in Re-Volt! Nuclear power has not been adapted to small installations operated by small groups or individual entrepreneurs. Nuclear power plants are run by large utilities and major power consumers. These facilities have outputs of from 300 megawatts to over

3000 megawatts. Such massive facilities require so many customers, or power users, that it is impractical to have them privately built. They are usually built by two or three major corporations and coordinated by government action. So far that has meant that essentially all nuclear power plants are government subsidized. That is not much different from the building of Hoover Dam or Grand Coulee dam. We must at least look at the nuclear situation and see how it can affect our change from large energy utilities to the small creative supplier.

There are already several designs for smaller community sized nuclear power plants supplying between 10 and 35 Megawatts. These newer designs have yet to be put on-line. They show much promise to make a better and safer use of nuclear power. It is no surprise that none of these is an American design. We have been so afraid of nuclear power for decades that we have halted all our efforts at improved nuclear power. In addition, these small plants would create lots of nuclear waste, and we have not put much real effort into how to manage and capitalize on nuclear waste. Until we do, the proliferation of small nuclear power plants should be undertaken only with extreme caution.

The nuclear accident at Three Mile Island terrified about 2 million people, but hurt none of them. We have learned from that incident. That is a soluble problem, and we have made some progress in solving it. If you live near a nuclear power plant, you have been approached about emergency evacuation plans. Let's not allow irrational fear to prevent us from taking positive corrective action. Coal, on the other hand has killed thousands of people. U.S. deaths in coal mining have decreased from a maximum of 3242 in 1907 to 451 in 1950 to 141 in 1990 and 32 in 2005. Oil casualties have been rising from about 70 in 2000 to 125 in 2006. Just from a safety point of view, nuclear energy is a far better option.

All of these nuclear materials are made by a refining process that concentrates uranium and plutonium-the elements that can undergo fission. The ores from which they are concentrated contain less than 1% of Uranium. (There are no plutonium ores, because plutonium is not a naturally occurring element. It is man-made.)The process is complex, and produces enormous amounts of waste--about 300 times as much spent ore as the fuel or weapons grade material they manufacture. The plants that make these are large, expensive and very sophisticated. The International Atomic Energy Commission (IAEC) monitors such facilities all over the world. (Not particularly effectively, considering the problems we have in Iran and North Korea.) As you can see in your newspaper, there is much international political intrigue over these facilities and their output. These materials are so hazardous that the purpose of the IAEC was to prevent further expansion of nuclear manufacturing, and hopefully eventually to get rid of weapons grade altogether. That seems highly unlikely at the present time.

It is unrealistic to declare that nuclear power is non-polluting. It produces radioactive waste that is also a subject of much controversy and debate. These waste products are primarily radioactive hazard, rather than a risk of nuclear explosion. Another pollutant common to nuclear power plants is huge amounts of excess heat, which strains nature's ability to dissipate it. The now closed Hannaford nuclear power plant produced so much excess heat that it warmed the entire Columbia River almost 8 degrees Fahrenheit until we took steps to control that. There is no new technology that doesn't create special problems. These can be managed if we all work together.

Here is another scenario. What can be concentrated can be diluted. We could establish manufacturing plants that dilute weapons grade materials to fuel grade material. This would reduce the potential for nuclear explosion, while supplying fuel to the 400 existing nuclear power plants and perhaps to some new ones as well. The threat of nuclear holocaust is magnified by the psychology of this ultimate weapon. Egotistical leaders of small countries that want to become international powers covet nuclear power, and particularly nuclear weapons. They know they will never use them because there are enough nuclear weapons spread around the world that any country that might venture such a move could be radioactive toast in about 2 hours. If we actually used only 2% of the nuclear weapons that exist today, the world would become so radioactive that nothing larger than a mosquito would survive. Nevertheless, they don't want to feel pushed around by nuclear powers. There is an enormous amount of ego and politics in nuclear power.

Jeremy Gorman

The United States has more than 100 times as many nuclear weapons as they could possibly use. We battle with Russia over who should dismantle what nuclear weapons and when. I propose that we, as a nation, unilaterally begin a slow (100 years?) process of downgrading nuclear warheads into nuclear fuel for nuclear power plants all over the world. We could become the major supplier of nuclear fuels to the 400 existing nuclear power plants, and encourage the building of more of them--small, modernized, safe ones. We might even make a little money doing it. Far more important is the fact that we would send a notice to the whole world that being a member of the nuclear club just isn't worthwhile. The expense is enormous, and the gain is non-existent. We were the first nuclear power in the world. We should show our intent by becoming the first nation to withdraw from the nuclear club. That message by itself would make a huge change in the attitude of the whole world. If we were to start such a program it would not be very long before Russia and other nuclear nations followed suit. Set a pattern of active nuclear disarmament. Even if it took twenty or thirty years for other nations to follow suit, we would have no handicap at all. Remember, exploding 2% of the world's nuclear warheads would make the world uninhabitable. Let's show the world that we really don't want to be in that nuclear club. The rest of the world would at least give it some serious thought. In the process we would establish a large and profitable industry and employ many thousands of people. This might be a carrot for the aspirant nations who want to be members of the nuclear club. Instead of enormous expense for a precarious political gain, they could make some money and be loved by their neighbors.

Let's not stop there. Nuclear waste is a growing problem and we spend many hours of national debate on what to do with those radioactive waste products produced in our nuclear power plants. There are at least two kinds of nuclear waste—high level and low level. Surprisingly the low level is much more hazardous than the high level. It cannot cause a nuclear explosion, but radiation makes low-level waste a serious hazard. Low-level waste has a short half-life—generally from a few minutes to about 100 years.

High-level waste, on the other hand, is a hazard because it contains materials capable of going through nuclear fission and producing a nuclear explosion. These waste materials have low concentrations and cannot explode unless they are re-refined and concentrated—an expensive process usually deemed unprofitable. They have long half-lives in the range of 100 to 20,000 years. Although it has a low level radioactivity it can damage people and animals that are exposed to it. Since they cannot see it, or hear it, or feel it, it is particularly frightening to the everyday citizen.

Low level radioactive materials can neither explode or be used in power plants. Many of them can find use in radioactive instruments and devices, such as medical examinations and treatment. Most, however, are a serious disposal hazard. Instead of letting them cause major radiation problems all over the world, let's consider a dilution process. Some 90% of the extreme radiation hazard has short half lives of under a year. It dissipates itself within two years leaving a serious but manageable radioactive material. We store those wastes on site at nuclear power plants for two years. We should then dilute that two-year old low-level waste with the depleted ores from which they were originally derived. Bring them to the radiation level of the ores from which they were originally derived. Spread them around the mines from which they originally came along with some young trees. This is a nuclear mine reclamation project. It will shorten the hazardous radioactive life of those ores. It shouldn't require special legislation or public debate. (Of course it will, but that is part of life.)It will not require the purchase of more land, because the refiners already own mining rights to the land which is unpopulated and in remote areas.

Let's also consider diluting weapons grade materials to fuel grade materials by combining it with high-level waste. We will not only have averted a nuclear power struggle, we will have produced a source of energy without greenhouse gases and solved a nuclear waste problem. If we could convince Iranian president Mahmoud Ahmadinejad to develop that project, he could also employ Iranians and reduce their growing poverty rolls and their increasing incursion into Iraq and Turkey and Syria. It might even re-establish our construction of new nuclear power plants—a process that has been stalled for forty years. We (the IAEC) would have to monitor carefully all nuclear shipments both to Iran and within Iran. They must assure that Iran does not gain control of massive amounts of weapons grade materials with which to make bombs. Even a megalomaniac like Ahmadinejad might like to be an international beloved hero and energy supplier rather than a hated adversary and enemy. He can employ thousands of Iranians, as he originally promised and rebuild his popularity among unemployed Iranians who have lost faith in him. Could Jimmy Carter or Nelson Mandela convince Ahmadinejad to go that route? How about Kim Jong Il in North Korea? Could Mikhail Gorbachiev convince him that North Korea would be better as a world supplier of energy than a world disruptor? Can this be an opportunity to convert an adversary into an ally? It surely is worth the effort.

Here is an idea that has not even been tried. Nuclear power plants produce massive amounts of extra heat and we have to take special care to prevent that heat from overheating local streams and lakes. So far that is all wasted energy. But energy is energy, and we use huge amounts of it. Consider the thermoelectric effect. This phenomenon is what makes thermocouples work. It is the ability to convert heat directly to electricity. There are thermoelectric coolers in use today. Why not put a huge number of them on the cooling systems of our smaller nuclear power plants, and increase their electric output while preventing damage from overheating local environments? I suspect we could increase the energy output of a nuclear power plant by 20 to 40% with such a scheme. But there is another facet to this. Nuclear power plants produce high voltage alternating current for the grid. The thermoelectric coolers produce very low voltage direct current. Here is an idea reinforcing the "alternate" in alternate energy. There are many uses for low voltage electricity, which might be supplied by this thermoelectric cooling system to make use of this waste energy and help protect our environment at the same time. Has anyone even considered such a program? I suspect it would not be a cheap solution, but such plants are designed to last at least thirty years and hopefully fifty years.

Nuclear waste is a cultural mindset. Waste is material that you can't use. Very often different people or operations want and need that same material. To them, it is not waste. The massive heat put out by the Hanaford nuclear power plant. It was a problem heating the Columbia River, but there are hundreds of uses for that huge amount of heat. Why throw it away, when we need heat in so many operations, including heating your house or hot water. Waste is a frame of mind more than it is a problem. I already mentioned that the electrolytic process for making hydrogen is inefficient. That is largely because we throw away most of its output—the oxygen we make at the same time. Oxygen is a commercial product which is in fairly high demand. Why waste it? It is another man's raw material. Many radioactive materials created in a nuclear power plant are valuable resources for other industries. Much of that material is not waste. It is just not needed in the current operation. Careful planning will allow us to make use of enormous amounts of "waste" materials, which have value in different industries. Our energy industry creates massive amounts of "waste", substantial, portions of which are useful for other purposes. We ignore them and focus exclusively on the energy we are trying to make. Waste is a cultural concept.

Many great ideas in energy either get stalled, or get sidetracked into a specific channel and begin to drift off the original purpose. Unlike many of the ideas in this book, nuclear energy is not one that you can practice in your back yard. But we can refine, shrink and improve them. Energy is the opportunity for ideas to be discussed, modified and improved. Remember, it is easy to find fault, but difficult to find solutions. Energy can and will grow by the blending and modification of good ideas with even better ones. The world of energy is waiting for YOUR ideas.

Chapter 8

Hijacked Hydrogen

Hydrogen is a conundrum. It is the most abundant material in the universe, but pure hydrogen does not occur naturally on earth. It is the most energetic fuel, but it is so light that it is extremely difficult to transport. The ore from which hydrogen can be made on earth is water. It is easy to make, but difficult to store. The most desirable production process, electrolysis, is only 70% efficient. Since electrolysis also makes oxygen, the fourth largest commercial chemical in the U.S., selling the oxygen would make the process cost competitive. Hydrogen is clean burning, producing only water when it does. It is extremely energetic, releasing two and a half times as much energy per pound as natural gas or gasoline. Despite all these self-contradictory properties, it has enormous potential in our energy profligate world. The science to utilize hydrogen has been known for centuries, but we have very little technology to capitalize on that science. Hydrogen is a challenge—a worthy one.

Although hydrogen is clearly one of the major energy resources of the future, our approach to this opportunity so far has been slow, expensive and not very creative. The challenge of a totally new energy source has somehow been relegated to current energy suppliers, and they have moved in a direction that will keep them in control. One of the biggest reasons for using hydrogen is that it is non-polluting. Currently 94% of our hydrogen is made from methane and water. This is an energy consuming process that produces four times as much poisonous carbon monoxide as it does hydrogen. That's non-polluting? And burning that carbon monoxide will produce 7 times as much carbon dioxide (our most prevalent greenhouse gas) as hydrogen. Sure, burning hydrogen creates no greenhouse gases, but you've already produced the greenhouse gases before burning that hydrogen. We make only 4% of our hydrogen by electrolysis of water. That process is not preferred because it is not very efficient (read less profitable). In addition, so far, electrolysis uses power primarily from the grid, 48% of which comes from coal—our most polluting fuel. So we are being misled and misinformed by those who are making hydrogen today.

Another neglected asset of hydrogen is that it is everywhere. Eleven percent of water is hydrogen. Where there is no water, there are no people either. (Take that with a grain of salt—southern California has almost no water, but is densely populated. They have been stealing water from their neighbors for over 100 years.) Why do they make hydrogen from methane? Because the current energy companies have lots of methane and can make a profit on it. Ask them what they do, with all that carbon monoxide or carbon dioxide that process produces.

That is not the only thoughtless diversion from our objective. Hydrogen is extremely light, weighing one fourteenth as much as air. Even at dangerously high pressures, it is difficult to store and transport. This problem is the largest single barrier to widespread hydrogen usage today. Why, then, do we direct most of our work on hydrogen into transportation, where the storage and transportation problems are maximized? For your furnace, your stove or your hot water heater, hydrogen might require some large storage tanks, but you don't have to drag them around with you as you do for your car. You either take a trailer along to carry the hydrogen, or put it in a tank under 15,000 psi pressure. California shows lovely pictures of the Governor filling his hydrogen vehicle at a gas pump—at 15,000 psi? The very least they could do is to pre-fill small (40 lbs.) portable tanks which could be purchased at the filling station in exchange for your empty ones. Forty pounds of hydrogen yields as much energy as 20 gallons of gasoline. However, you would need a mighty strong and heavy tank to hold 15,000 psi.

Jeremy Gorman

Your gas stove, your gas furnace or your gas hot water heater can be converted to burn hydrogen for a nozzle change costing less than $20. Your car requires a $50,000 fuel cell. Since hydrogen currently costs more per energy content than gasoline, what's the incentive to switch to hydrogen-powered autos and trucks?

Hydrogen is an ideal source for future energy. It is our approach to hydrogen usage that is off base. The idea is great, but we haven't thought it through. Like all energy sources, hydrogen has problems. Solutions exist, but we seem to avoid them rather than solve them. The science is known, but the technology to capitalize on it has been neglected. We have been applying coal, oil or gas solutions to hydrogen unsuccessfully. If we continue on this path they never will be successful. Let's look at these barriers and learn how to add hydrogen to our energy sources and relieve some of the pressure on coal, oil and gas.

There are six primary problems:
1. Hydrogen does not occur naturally on earth—it has to be made.
2. Hydrogen's extreme light weight and low boiling point are the major barriers to its common usage.
3. Hydrogen burns with an invisible flame, and consequently presents a safety hazard.
4. Hydrogen is colorless, odorless and tasteless, so it can be present in dangerous amounts unknown to the user.
5. Hydrogen causes embrittlement of most ferrous metals, and makes them subject to failure under shock or stress.
6. Hydrogen cannot stay in the earth's atmosphere, so any spills will be lost forever.

Only one of these is a fatal flaw, but our world has not systematically addressed any of them. That is why hydrogen is not in common usage anywhere in the world. It should be and, if we do our homework, it will be. Will that be in 5 years, or 150 years? That is up to us. Do we have 150 years to solve these problems?

Let's start with problem number one. Since hydrogen is the most abundant element in the universe, why doesn't it occur on earth in free form? The earth is warm. Temperature is the measure of the average kinetic energy of the moving molecules of the substance being measured.

Kinetic energy is a product of the mass of any object and its velocity. That means that light molecules are moving faster than the heavy molecules at any given temperature. Since hydrogen is the lightest molecule it is moving faster than other molecules. On our warm earth, hydrogen molecules are moving faster than the escape velocity from our earth's gravity. So hydrogen spilled into our atmosphere will escape into outer space, and be lost forever.

So there is no natural hydrogen gas on earth. There is, however, an abundance of hydrogen compounds and most particularly water. There is three hundred times as much water on earth as there is air. Hydrogen can easily be made from water. There are several such processes. One is by electrolysis—the electrical process that breaks water down into its two components—hydrogen and oxygen. This is an energy absorbing process that has no other by-products. You get no pollutants. You do get eight times as much oxygen as hydrogen, and oxygen has commercial value.

Electrolysis is not the process favored by industry. It is slow and less energy efficient than the methane process. Energy companies have an abundance of methane gas and see the manufacture of hydrogen from it as another source of profit. They are correct in noting that it takes less energy (and less expense) to make hydrogen from methane and water than from water alone. Getting down to the basics, however, hydrogen is not a *source* of energy because it does not occur naturally like coal, oil or natural gas. Hydrogen is a *conduit* for energy. You get out what you put in. In electrolysis, you put in the electric energy, and get the hydrogen for diverse uses, without making any greenhouse gases or other pollutants. From methane you get more hydrogen for your energy input, but you make four to seven times as much greenhouse gas as you make hydrogen and you get no oxygen.

The electrolysis process becomes tainted when you use power from the grid. Truly clean energy comes from solar, wind or water power. Viable hydrogen manufacturing processes for each of these are in use today, but in miniscule amounts primarily at universities. The objective of industry has been to make money, not to make clean energy. The clean energy concept of hydrogen has been hijacked by the existing energy industries to serve themselves, not the public. To live up to its potential as a clean energy source, hydrogen should be made by electrolysis using electricity made from clean sources like wind, water or solar power. We must look at the total process and not be misled by considering only a portion of the whole process. We are touting the future use of hydrogen fuel cells in transportation. It would be far more efficient and cheaper to use that methane directly in our existing automobile engines. We already have small fleets of trucks and school buses running on propane. These vehicles can run on methane with a minor modification rather than a very expensive fuel cell. Let's face it. The fuel cell is a scientific wonder, but is impractical for widespread use.

Problem number two is the extreme light weight of hydrogen. To get a feel for the size of the problem, consider a propane tank that you use in your outdoor grill. It holds twenty pounds of propane at about 50 psi pressure. That can produce 400,000 Btu of heat. (Btu, or British thermal unit, is the amount of heat required to heat one pound of water one degree Fahrenheit.) To get that same heat, you would need only 5.7 lbs. of hydrogen. But at 50 psi that tank will hold only 0.008 lbs. of hydrogen. It would take 735 such tanks of hydrogen to produce as much heat as the propane. An obvious answer is to compress the hydrogen. You need 35,000 psi to put that much hydrogen into such a tank. You'll need a mighty strong and heavy tank! You'll also need a powerful and expensive compressor.

To overcome that problem, industry has considered special hydrogen adsorbing materials to contain the hydrogen. Lithium hydride and carbon nanotubes have each been found to be effective. The expensive lithium hydride holds 6% of its weight of hydrogen. The even more expensive carbon nanotubes will hold 8%. Our 18 lb. propane tank would hold 32.5 lbs. of lithium hydride and 1.95 lbs. of hydrogen. Instead of weighing its current 38 lbs., that tank would weigh 54 lbs., and you will need three of them. How do you coax the hydrogen out of the hydride? Not just by turning a valve.

It is clear that storage and transportation are the biggest challenges to everyday hydrogen usage. Let's consider an obvious option. Can we dissolve hydrogen in other liquid fuels and enrich them? Very much like dissolving carbon dioxide into drinking water to make seltzer water, hydrogen may be soluble enough in some fuels to reduce their greenhouse gas emissions by 30% to 70% while nearly doubling their heat output. One of the best features is that you could continue to use existing fuel transportation and storage facilities with almost no modification. You would no longer need massive compressors or heavy duty storage tanks. I have been on a four-year campaign to at least try this, but have had no response whatsoever. So far they have been more interested in making money than in solving the problem. (There are more details on this possibility in a later chapter.)

The third problem, the invisible flame, is easily soluble. Add about 1/2 % of ethylene to the hydrogen, and the flame will be yellow and visible. Sure, you make a tiny amount of carbon dioxide, but about $1/500^{th}$ of the current emissions.

The fourth problem, the colorless and odorless character of hydrogen flames, is also easily soluble. Add about 0.02% of skunk oil to your hydrogen and people will soon know it is there. This may also help solve the third problem, since the skunk oil will probably color the flame as well. It will also produce minute amounts of toxic sulfur oxides.

The fifth problem, hydrogen embrittlement of the container metals, is indeed a challenge, but has already been solved. There exist today many alloys that do not suffer hydrogen embrittlement. But there is also the possibility of using filament-wound fiberglass tanks. They are lighter and stronger than metal tanks, but beware! They fail by explosion, not by cracking and leaking.

The sixth problem, the escape of hydrogen from earth, has no current solution. The abundance of hydrogen resources makes this an unlikely barrier to hydrogen usage, but profligate mankind has a way of overstraining our most abundant resources, like the forests of the earth and the massive stores of coal and oil. We are assuredly capable of overstraining the hydrogen resource too. Although extremely remote, even with massive use of hydrogen as a fuel, the depletion of the world's hydrogen by accidental spills into the atmosphere would be a permanent disaster, because we cannot replace it. If you would like to live on hot, dry Venus, that is your option. It *can* happen here.

Considering the huge potential for hydrogen, I would like to establish a "Dual Fuel" industry that specializes in blending hydrogen into other fuels. Here is a way of capitalizing on the advantages of hydrogen using equipment already in commercial use. Small tanks that hold 20 to 500 lbs of liquid and gaseous fuels are already commercially available and in common usage. Existing compressors can add hydrogen and boost their output while simultaneously reducing greenhouse gas emissions. Unlike the lithium hydride and carbon nanotube systems, you don't have to ship the expensive absorbing agent back to the point of origin for refilling. The absorbent is low cost fuel and is consumed on site. It would not be much of a challenge to place this capability in your home and capitalize on your solar or wind generated hydrogen.

Here is an opportunity for all you creative readers. Why don't you start a small company whose objective is to find the optimum fuel hydrogen blend and sell it to a demanding public? There is certainly no shortage of customers, and you could learn and grow as your technology gets ever more productive. You would also create many jobs for our underemployed economy. The science is known, but needs to be focused and refined. This is the kind of project that lends itself to the small local entrepreneur, rather than Big Oil, whose main concern is to enhance its bottom line. Are you up to that challenge?

Chapter 9

Ossified Electricity

Electricity is not a source of energy. It is, however, probably the most prevalent carrier of energy in the world today. It has a long history of remarkably creative developments that allow it to be so prevalent and so necessary in our lives today. But this country, and maybe the whole world, has a mindset about electricity that is beginning to cause us problems. We are straining our supply, and we need to take a look at it and see if there is another way to make use of it. We are getting carried away without looking closely at what we are doing.

Edison's light bulb worked on Direct Current (DC). Direct current is the flow of electrons from a negatively charged source to a positively charged receiver. Electricity has a shipping problem--it dissipates itself along the way because of the friction (resistance) in the wires, or transmission lines. In order to make a light bulb work, you need a substantial flow of electricity called amperage. However, if the light bulb or television set is a long way from the source, the resistance in those wires depletes the electricity rather quickly (Line Losses). Even the best electrical conductors, silver and copper and aluminum, have enough resistance that you cannot send direct current from New York to Washington, DC, because the friction would eat up all the energy and there would be nothing left in Washington. (There are people who think that would not be such a bad idea considering the politicians there.)

Nicola Tesla noted that those line losses were exclusively dependent upon the amperage or flow---not on the force, or voltage. Since power is the product of amperage times the voltage, he used Michael Faraday's "transformer" to create extremely high voltages, so that power could be transmitted long distances with very low amperage and small line losses. However, transformers won't work with direct current (DC), so alternating current (AC) is required. Hence the huge towers carrying "High Tension" wires with 345,000 volts of electricity. On them you can send electricity from New York to Washington at far less, but still substantial, line loss. There are currently efforts to further increase the voltage to 765,000 or 1,100,000 volts to minimize what have become substantial line losses. So remember that the prime reason that we use AC is so that we can send electricity long distances without dissipating it in the transmission lines. It eases a transportation problem. But you can't store AC, so you have created a storage problem.

But what if we want to use the electricity right where it is generated, like in a solar panel or a windmill on your rooftop? A solar panel can't make alternating current. (There are devices called "alternators" that can convert DC to AC at considerable expense and some loss of energy.) That merely means that we can't send solar generated electricity over long distances. But you don't need AC to send electricity from your roof to your living room.

Now we have a more recent remarkable invention. It is called the LED (Light Emitting Diode). It has two great features—it generates two to five times as much light from the same energy as fluorescent or incandescent lights, and it operates on low voltage. Why not use solar or windmill electricity to light your house with LED's? You'll have low voltage, but short wires. You won't have to pay Con Edison for your electricity. Currently the barrier is that LED's are expensive and not cost effective. That will change as LED's become mass produced and more reasonably priced.

In the 1920's and 1930's, few houses had electric wiring, and almost none outside of big cities did. That was a period of Rural Electrification. It was a system that sent electricity to homes, stores, factories and offices nationwide from a relatively few major electric power plants scattered around the country. That was a marvelous idea, but it made us all think that electricity comes from "The Grid" – three regional systems of electric power transmitted over high tension wires. Today it does. Because most homes were long distances from electric power plants, AC and the Grid completely dominated electric power. We essentially abandoned DC as a source of electric power--except in the automobile. The automobile needs to store electricity so it can start by itself no matter where it is. You can't store AC!

We need Rural Re-electrification today. We are creating a need for two separate electric systems. One for the burgeoning low voltage demand that requires the ability to store electricity, and another for the major power consumers that are not located near a power station. It is a simple process to add wires to carry DC power within the home or office. It is less of a problem, than was the Rural Electrification, since the hardware already is commercially available. The whole idea of alternative sources of energy can be easily accomplished in the home or office with a simple second wiring system. Most homes should have both a high voltage AC set of wires sourced by the Grid, and a low voltage DC set of wires sourced locally. That is cheaper and more energy efficient than the system of adding local DC power to The Grid, as many do today at considerable expense and substantial loss of energy. That is not a desirable plan. The Grid encounters variable demand. It increases at night when you turn on the lights and when DC power from Photovoltaic panels is not available when no sun shines. Since you can't store AC, that means you add to the grid when it least needs it, and can't add to it when demand is highest. That is not a good deal for either the electric company or the user.

There are more and better reasons for two electric systems. In our modern electronic age, more and more power is used at low voltages. Computers run in 5 volts. Automobiles run on 6, or 12 volts. Flat screen computers and television sets have converted from the high voltage cathode ray tube to low voltage digital flat screens. More and more electricity is going low voltage. As we miniaturize things, they need thinner and thinner insulators that fail under higher voltages. We are developing a two tier electric demand--high voltage for large power demands and for shipping long distances, and low voltage for everything else. It would make sense to have both a low voltage DC system and a high voltage AC system. DC appliances are already commercially available for motor homes, campers and boats.

Big Energy doesn't want you to think that way. They will come up with thousands of reasons why that is a bad idea. They are debating 50 billion dollar changes in the aging Grid to bring it up to date, meet increasing demand and, of course, to keep them in control. They will tell you that your lights won't work unless the sun is shining, or the wind is blowing. Ever heard of a battery? Unlike AC, we can store DC in a battery or a capacitor (also called a condenser), like you do in your automobile. When the sun shines or the wind blows, you can charge your batteries and use your LED lights or DC appliances whenever you need them. A major use of that local electricity will also be to charge your hybrid or electric automobile or truck.

There are other differences in AC and DC power systems. In DC you have a positive wire and a negative wire. They are always the same. If I connect any positive wire with another positive wire they will work together. But with alternating current, there is another problem. The voltage on an AC wire reverses 60 times every second. That's called 60cycle. (It might be worth noting that in Europe, the standard is 50 cycles per second.) Connecting two wires could cause problems if one source has a positive voltage when the other has a negative voltage. To make AC circuits work together, they have to be "in Phase". That means that each peaks at the exact same time and reverses on the same schedule. That is a primary purpose for "The Grid", which puts all of the AC in a region on one phase so they don't get out of phase and fight each other. There are special devices called phasers and "inverters", which assure that both wires are in the same phase before they are connected. Such "phasing" often causes substantial energy losses, and adds substantial cost to the system.

Commercial wind generators are designed to feed electricity onto the national Grid. It requires not only exactly 60 cycles AC, but the feed must be exactly "in phase" with the system. When windmills start up they not only have to be brought to exactly 60 cycles AC, but also have to go through a phasing process. Neither would be necessary if we generated DC, but the DC can't be sent long distances. That is not a problem within the home. The ideal future home will have two sets of wiring—an AC circuit for the grid and a DC circuit for locally generated power. As society becomes more advanced and more complex, alternatives will become a requirement.

This is a prime reason that there are not more wind generators. These limitations on wind generators explain why commercial windmills are very large, and won't operate at low wind speeds where they can't get enough power to add to "The Grid".

Since the grid is a (actually three) nationwide operation, this entire continent is on the same phase. Any additions to that grid must be phased to that cycle. The system will not tolerate any deviations. That is one of the problems they face when there are large power outages—all restoration must be exactly phased to the grid.

But we are missing the main value of wind or solar generation. Wind and sunshine are everywhere. If you generate electricity on your roof, you don't have to buy it from distant sources. You don't have to make AC. You don't have to phase the output. You can store it in batteries for use when the wind is not blowing, or when the sun is not shining if you use photovoltaic panels for electricity. You can run computers, cell phones, LED's and automobiles on low voltage DC. The DC markets are building. Shouldn't the supply increase as well?

Today the entire windmill industry is focused on The Grid, and we largely neglect several major sources of energy in this world. Let's all build a future where most of us have a solar panel and a windmill on the roof, which charge your hybrid car battery, and run your LED lights, your computer, cell phone, your flat screen TV and many small kitchen appliances. No, you will not replace the grid, but you will reduce your dependence on it, and cut your electric bill by from 20 to 80%. You will not have to spend 50 billion dollars to update and modernize the Grid.

The power companies will point out that you can make AC out of DC with special inverters, and phase it to their grid with other special equipment. And, if you make more than you use, they will reimburse you for the power you put on their grid with yet another device. But these are expensive devices and lose substantial power in their operation. Adding DC wiring to your house will cost less than that equipment.

In summary, then, wind, solar and waterpower will find their most practical applications in generating local DC for local use. They will not replace the Grid, but will supplement it. DC can potentially supply between 20% and 80% of household needs. It will probably require separate wiring—a small price to pay for so many advantages.

The emphasis again is bringing the sources down to the individual. Don't let Big Electricity and Big Oil dominate your energy usage. Do it yourself. Build a solar panel! Build a windmill. Put a water wheel in your local stream. There are instructions available on the Internet. Using them will not only save money, it might help save the world. What's more, if we all do it, it will create massive job opportunities and create thousands of small competitive businesses. Isn't that what we need since we shipped all our jobs overseas? We will also save the polar bears and the shrimp.

There are other advantages to dual electric power systems. If one fails, the other is still available. That would be a great boon during massive power outages that occur after earthquakes or major storms. And in addition, once you have installed an alternative electric source, it is free from then on. No wonder Con Edison doesn't like the idea.

But remember the payback factor. Solar panels and windmills are expensive today. They are built for the elite energy market—mostly for show. But if they were small and mass produced, they would not be so expensive to install. Putting a small windmill on your roof today costs around $3000. I've seen systems that cost over $12,000. Putting solar panels there could cost over $16,000. It would take a long time to repay that cost with the electricity they generate. (Most homes pay around $200 for electricity every month, and these devices probably will replace about half of that.) But they could and should be mass-produced in large quantities by small local companies to get the cost down. You should be able to put a windmill on your roof for about $400. It would be a low voltage system, but today's electronics and automotive industries are already demanding lower voltages, so that they can miniaturize. So think about it. The Energy Revolution, which is the subject of "ReVolt!" is the one that brings power to the people, not to profit-oriented corporations. It doesn't take a genius to do that because the science is already known. With help from the Internet and from your friends and neighbors, you can do it. You'll learn a lot and save a fortune. Why not give it a try?

Three ideas summarize this change. First is you, I, we. Not Big Oil, the Grid or Government. Next is Payback. This is not an exercise in social climbing, but a practical way of saving the planet and saving money at the same time. The third, and most important, is alternate. Every source of energy should have an alternate. That is the key to being in control of your future energy need, rather than a victim of it.

Chapter 10

Wasteful Water

We see lots of windmills, and much activity in developing more of them. That is wonderful, since wind is a huge source of solar energy. But consider this. There is much more water in the oceans than air in the atmosphere. The Sun evaporates that water, and sends it up into the atmosphere to form rain and rivers and streams all over the world. So, not only is there lots of water, but it is constantly moving. Since water is almost 800 times as dense as air, there is lots of useable energy in moving water.

The power of water was not lost on the ancients, who had devised water wheels long before the birth of Christ. The Egyptians, the Persians and the Romans used waterpower. There was much use of waterpower until coal was found and the steam engine invented. They were easier, so they gradually displaced waterpower, as Man demanded ever more energy. We use little waterpower today. Most of that is hydroelectric power that supplies about 5% of our national electric power. Except for Niagara Falls, we derive almost no power from natural waterfalls. We have not been very creative about our use of waterpower. Waterwheels are not exactly state-of-the-art technology. Water wheels drove flourmills for centuries. But it went beyond that. The first American use of waterpower to generate electricity was at Niagara Falls in 1886 when commercial electricity was still in its early infancy. It has been modified and improved dozens of times since then. Niagara Falls has been supplying electric energy to the United States and Canada ever since.

There is only one Niagara Falls. However, its example stimulated other ideas including the use of dams to capture waterpower. We had already built many dams to capture waterpower, but the first major use of water for electric power was at Hoover Dam. Built in the 1930's, it went on stream as an electric power source in 1941. Soon thereafter we built Grand Coulee dam in Washington, which is still the second largest waterpower facility in the world and the largest concrete structure ever built. We now have 2400 hydroelectric dams generating electricity in this country. These are good and creative uses of our natural resources. Almost without exception they were massive projects built or at least subsidized by taxpayer money. They are operated by profit-oriented organizations that feed electricity onto The Grid from which you buy your electricity. Because of the taxpayer subsidy, they were kept under strict regulatory control to assure that the public got what it paid for.

This system has indeed helped produce our national electric grid upon which this nation depends heavily for essentially all of our electric power. But it has also narrowed our thinking on power. We are still using technologies developed 100 years ago, and they remain in the control of a few large profit-driven corporations. It is time to think outside the box and find new ways to capitalize on natural resources.

So far almost all our waterpower is from falling water from dams, or waterfalls. But water flows steadily in rivers and streams, as well as in ocean currents. Unlike wind, water flows fairly steadily year-round with far less variation than wind. A windmill is dependent upon one of the most fickle sources of energy in the world. Not so ocean currents that flow in predicted and mapped patterns 365 days per year. Rivers and streams are much more consistent in their flow patterns than wind. But aside from massive dams, we use almost none of that energy. Why do we build such massive installations? Primarily because when governments subsidize projects, they want to serve as many people as possible, so they build huge structures. Most of the dams and waterpower plants in this country, as well as in the rest of the world, were built either by a government, or with a government subsidy.

We should consider building thousands of small water turbines that could be used in ocean currents, rivers, streams and tidal basins to generate energy all around the world. (See the chapters on Tides and Ocean Currents.) These resources contain enormous amounts of unused energy that is readily available with relatively simple devices that are already on the market, and cost less than your kitchen stove. They can be located anywhere that water moves, so it need not be sent hundreds of miles to the user. Consolidated Edison doesn't want you to do that, because it will cut into their sales and cash flow. It will help snatch energy from their control.

We have the technology and even the hardware to make use of tides, streams and ocean currents. They are almost completely absent from our energy sources. Why? The main reason appears to be that they do not adapt to massive installations on which large corporations can make money. These are people resources that will ultimately be run by small, local enterprises and individual entrepreneurs.

But ocean currents and streams are not the only sources of waterpower. Water also is a source of lunar power. Twice each day, billions of tons of tidal water rise and fall all around the world, invading and deserting our coastlines everywhere. Many of those coastlines already have breakwaters to control that water flow, and to protect them from erosion. It would not be such a huge task to install small water turbines every few feet along those breakwaters. As the tide flows in, the wheels would generate electricity by turning in one direction and, of course, generate more power turning in the opposite direction as the tides recede six hours later. The mass of water is so great that vast amounts of energy can be produced at no cost beyond the installation cost. You don't have to pay Con Ed. There is a whole chapter on tidal energy later in *Re-Volt!!*

Water lends itself to creative thinking. These are just a few of the hundreds of ways water can supply useable energy. It need not always be electric power. Suppose we used that power to generate hydrogen—a clean and highly efficient medium for storing energy. Such hydrogen could be used in kitchen stoves, hot water heaters, home heating furnaces and even in automobiles. Its extreme light weight and low boiling point make storage and transportation the major impediment to its use.

Ocean currents lend themselves to large installations. I'm not sure if ocean currents are solar power, or lunar power, or perhaps they are a combination of those and the Coriolus effect. The Coriolus effect is powerful but little understood. Except on the equator, movement either north or south involves moving closer to or further from the earth's axis. You don't notice it, but on the surface of the earth at the equator you are moving eastward at 1000 miles power hour. But at Miami, Florida, you are moving eastward at only 900 miles per hour because you are closer to the earth's axis and don't have as far to go to go around the axis. At New York, you are only moving at about 760 miles per hour, and at Vancouver, only about 670 miles per hour. So, as you move north you will drift eastward because you are covering ground that is rotating slower than you are because it is closer to the earth's axis. In the southern hemisphere, as you go south the same thing happens. If you go north in the southern hemisphere, you will drift westward, because you will be coming to land that is moving East faster than you are. That is why whirlpools in the north go clockwise, while those in the southern hemisphere go counterclockwise. Since they involve rising air, storms (e.g. Hurricanes) rotate counterclockwise in the north and clockwise in the south (e.g. Typhoons). The Coriolus effect becomes ever more powerful the further you get from the equator, but is non–existent on the equator. It appears to be a major contributor to the force behind ocean currents, but not the only one.

Establishing an industry that builds, installs and services waterpower would employ thousands of people, and provide an alternate to the single source Grid. Next time a winter ice storm leaves you frozen for three days because there is no power to run your furnace, think about that alternative. The science of waterpower is well known, but the technology to capitalize on it lags behind the science. So far, most of it is for huge installations that are expensive and have long pay back periods. It took over twenty years to pay back the cost of Hoover Dam, and had it not been for inflation, it would have taken forty years. We already make small water wheels, which will generate small amounts of power. If our government is going to subsidize water power it would be far better to subsidize small creative companies who can mass-produce small units suitable for the individual homeowner or the small community. If they are close to the user, they need not be fed onto the grid where people have to pay for it. Here is a source of jobs to replace those we have sent overseas in the last two decades. These are manufacturing jobs. The Internet revolution has led us away from the manufacturing skills that were the mainstay of America for over a century. Energy can bring them back.

There is relatively little existent technology for water wheels. A few companies sell them as curiosities. They hold the potential to become major contributors to our energy supply. Here is an opportunity for ingenious people to start small businesses to capitalize on this neglected source of energy. The logistics of placing these water wheels is something of a challenge, but it is not insurmountable. Considering the rising cost of electricity it could be quite profitable for some creative entrepreneurs. We already have engineers who want to build another Hoover Dam, but the day of massive dams is over. They cause environmental problems, and we already have more dams than we need. Thousands of small water wheels in streams and rivers should be substantial contributors to our renewable energy era. That is a new twist on an already developed energy source that has become ossified and lost its creativity. Your ideas could create another source of renewable energy, save us all money and perhaps create a viable new business.

Many of these power devices are new and require minor design changes. Not all of these changes are completed and few are in use in remote communities scattered sparsely over the earth. This represents an opportunity for creative entrepreneurs to start small businesses that attack niches of power demand in myriad special circumstances. This is a challenge to our creative populace. These devices could be made in the United States and, if we do it right, could be exported to other countries and help balance our trade with other countries. Millions of people now live in energy-starved places in remote areas without the local resources to make their own power supplies. Creating special energy sources adapted to those local circumstances could be extremely beneficial to those communities and a source of income for our creative entrepreneurs. We are now an immense importer of foreign goods. Many of those are made by American companies overseas to capitalize on low labor rates. New technology made in the U.S. could be attractive to foreign countries. Those governments, like the United States, have tended to fund massive projects made by huge companies. Such projects never make it to remote villages where the need is greatest. The real opportunity is for our creative Americans to build hundreds of small companies to fill those niche needs in energy. In the last few decades we have fallen behind much of Europe and Asia in alternate energy. Now is the time for us to put our ideas into practice right here. It is also the time to create thousands of new jobs in a massive old industry that is crying for new approaches.

Anchoring many small water wheels in rivers and streams would not be very difficult. Small towns on rivers could establish local power companies making electricity for local consumption. There already are a few small dams that have little power stations in them to supply local users. I have already mentioned one in Searsburg, Vermont that generates a megawatt of electricity for the local power company. It is nearly eighty years old, and still works fine.

The very same technology could install water wheels in local breakwaters along coastlines everywhere. There is an enormous amount of energy available, but it has remained unused. It's about time we woke up and put it to good use. Federal stimulus money should be readily available for such projects. Why don't you get a local group together to study such opportunities in your town? Every town on a river should have one. Be a local hero, and start one in your town. The technology and some commercial equipment are readily available on the Internet. But it will require some careful planning. A 12 volt DC power station can serve a community of about 400 acres. If you want to serve a square mile of residents, you may need higher voltage, and perhaps AC to overcome the line losses in moving electricity more than a mile. That is not a problem, if we face it in advance and plan accordingly. Here is a golden opportunity for creative thinking. Is it your employment opportunity?

Chapter 11

Tides--A Neglected Resource

Not all our energy comes from the sun. The moon is also a source of power. The gravity of the moon, only one eightieth the mass of the earth but only 243,000 miles away is substantial. It literally squeezes the earth and changes its shape. These very regular lunar forces are called tides. There is a tide of about six to eight inches in the solid ground beneath your feet about twice every day. Actually there is a tide of about one inch from the sun's gravity every day as well. But by far the biggest tide is in the ocean, because water flows freely under the force of gravity--any gravity—the earth's, the moon's or the sun's. Tides vary greatly in size, but almost none in frequency. There is a slight variation on the timing of the tides because the sun and the moon go into and out of alignment about every seven days. That is a significant fact when you realize that it is that cycle that produced the week in all calendars all around the world despite enormous other differences in cultures. Tides are entirely predictable, and are in fact recorded in tide tables that are widely distributed and much used by ocean going vessels. But surprisingly little use of them is made for any other purpose.

One of our most neglected natural energy resources is lunar or tidal power. That is surprising since the energy in those tides is vastly greater than the air that gains so much attention as an energy resource. In addition, water is much less fickle than air. Twice each day, billions of tons of water rise and fall all around the world, invading and deserting about 60,000 miles of coastline everywhere. Many of those coastlines already have breakwaters to control that water flow and to protect them from erosion. It would not be such a huge task to install small (two to five feet in diameter) water turbines every few feet along those breakwaters. As the tide flows in, the wheels could generate electricity by turning in one direction and, of course, generate more power turning in the opposite direction as the tides recede six hours later. The mass of water is so great that vast amounts of energy can be produced at no cost beyond the installation cost. You don't have to pay Consolidated Edison.

Our approach to this massive energy supply has been typical of our approach to widely distributed power sources. Build huge installations with public money and charge the public to recover the cost. As a result, there is only one major tidal power plant today. It is the 240 Megawatt tidal power plant on the Rance River in France. It has been operated by Electricite De France (EDF) since it opened in 1966, and produces 600 billion Watt Hours of electricity every year. It has long since repaid its very high cost of construction.

Our ossified thinking on tidal power has stultified development of this vast renewable energy source. Small water wheels are already made in sizes from about 6 inches to about twelve feet in diameter. They are not very common, but they are commercially available. These wheels will work in tidal breakwaters as well as in small streams and rivers. We need not build Hoover Dam to gain power from flowing water. The idea behind Hoover dam was to make a massive energy supply that would supply thousands of homes that had no electricity. It succeeded and powered many factories as well. To pay for the costs of construction of the dam that was built by the government during the depression they charged 4 cents per kilowatt hour for the electric output. "Rural Electrification" was the by-word in the 1930's. Not today! Perhaps it should be. Why not have thousands of small generators, which are designed to supply local houses, buildings and communities? They need not supply all the electricity for those homes, but should supply some of the energy for most of them. There should be lots of local projects to build small dams or sluiceways, or just to submerge water-driven generators in streams or tidal basins for local use. There would be essentially no environmental impact. It will require local ingenuity, and provide lots of local jobs. Cooperation and involvement are the keys. You could initiate such a local program and meet some very creative neighbors in the process. In addition, American workers being as creative as they are, you can bet that there will be new developments in the hardware to make it more efficient and reliable.

Our use of other sources of energy to make electricity has been dominated by large industries that make electricity and want to keep control of it. They modify their equipment to make 60 cycles AC for the Grid, so that electricity can be sent over long distances. But electrical demand is going toward ever-lower voltages as miniaturization dominates the electronics industry. The phased 60-cycle AC demand not only requires the use of expensive conversion equipment, but also limits the time during which small local sources are able to meet their strict requirements. If we used energy where we find it and generate electricity for local use in low voltages, we can open up the production process to thousands of small suppliers. Let the grid send electricity long distances. The dominant electrical industry is trapped in the current system and opposes the dispersion of energy generation among the users. They pay lobbyists to persuade legislators to keep them in control of the power supply. If we do our homework, that should change in the next two decades.

There are many communities that were built on tidal basins. Primarily fishing villages, these towns have access to vast amounts of tidal energy. It is there that many of our tidal breakwaters were built to control and protect fishing fleets and local fisheries. Since these are usually close knit communities, they could and should get together and construct tidal generators in their local breakwaters. Since we have overfished much of the world's ocean, many of those towns have high unemployment rates. Why not divert some of those unemployed to the creation of local power companies to make the town energy self-sufficient? These could be a municipal source of electricity or perhaps hydrogen. These systems can be accurately scheduled, because tides are so well known that we can predict not only when tides occur, but their size as well. We have already predicted them months and even years in advance. Scheduling electric output would be both simple and accurate.

Tides are not confined to tidal basins and river mouths. They occur wherever ocean meets land. Creative people can establish tidal power mills on essentially any shoreline. If they and their neighbors are close to that shoreline, tides would probably be an excellent source of alternate power—one that is free after the cost of installation is recovered. These need not be massive installations like the Rance River station. Water is so dense that payback periods in all probability would be quite low (less than three years). And depending upon the local conditions, the output need not be confined to electricity. They might even consider generating hydrogen with that electricity so that the energy could be more broadly dispersed. It could help operate local furnaces, hot water heaters, stoves and even hybrid or electric automobiles. Build a local hybrid or electric car charging station for people who don't have wind or solar facilities at home. If your coastal town is widely dispersed, hydrogen may be a better way to move the tidal energy around than a local wiring system. If you live in such a community, organize an energy committee, and find out what best suits the needs and capacities of your town. Information is available at http://www.edf.fr/html/en/decouvertes/voyage/usine/retour-usine.html

That does not rule out a few massive tidal generators like the one in France. There probably are 20 places in continental U.S. where such large installations would be a valuable resource. Perhaps a call to EDF in France would help you make your plans.

Perhaps you have a better idea. We all could use more new ideas, and here is a way to encourage and capitalize on them. Small waterwheels already exist. Compared to wind, the variation between maximum and minimum natural water flow is at least an order of magnitude less. That means that waterwheel designs would be more uniform in both design and function. They are less complicated and technologically demanding to develop. Since many such designs already exist, there is a wealth of data already available for the massive expansion of waterpower. That doesn't mean that there isn't great opportunity for the further advance of this greatly underused resource. The creativity will come in the form of small nuances that apply to specific sources of waterpower. We are a creative nation, whose creative output in water power has been sadly lacking for over fifty years. Perhaps *you* can change all that. Good luck!

Chapter 12

Ocean Currents

The massive amount of water in our oceans is in constant motion. Ocean currents are strong, steady and remarkably consistent. They are already mapped. Here is a remarkably large source of energy that has to date been totally ignored. Why does man spend so much effort taming the elusive and fickle wind while completely neglecting a much larger energy source that is neither fickle nor elusive? The answer, of course, is in your face. How often have you been attacked by an ocean current like you were by a wind that blew your hat off? Few people have any intimate knowledge of ocean currents.

But the energy is there waiting for our creative souls to capture and use it. We have the technology to utilize that energy already in production. A variety of water wheels are already on the market, although as far as I know, none has been used to capture ocean currents commercially. All that is required is a little planning. We don't even have to get a permit or buy land to plant a waterwheel in the ocean. Of course you need to stay out of shipping lanes and find a way to get the energy back to shore where we want use it. Two relatively easy methods already exist. Generate electricity and put wires on the generator, or generate hydrogen and send that ashore by pipe or in a container. Unlike many other alternate energy resources the hardware is already in production and on the market. All we need is a plan to organize them into an effective package. Once they are assembled into a power station, they will produce power steadily for fifty to one hundred years.

Consider this. The power of ocean currents is greater than all the wind in the world. But we don't have single power plant of any size deriving its power from ocean current. Why? Unlike wind towers, you don't have to buy land for an ocean water wheel. You don't have to fight NIMBY battles with the local populace, a process that delays wind projects for years with punitive legal costs. There is no reason that they have to be massive projects. Thousands of small water generators (500 watts to 10,000 watts) can be placed in hundreds of ocean currents where ocean traffic is minimal or absent. In addition, they can be put in local rivers and streams so that they can easily generate electricity or hydrogen for local use. Water wheel generator designs already exist. Some are on the market. We don't use them because we lack the organization to do so. Also, power utilities oppose any such action that will dissipate their control of energy.

Distance is not always a problem. Some ocean currents are very close to shore, like the Gulfstream that runs within 5 miles of the of Florida coast for about 100 miles. I suspect that the city of Fort Lauderdale could derive almost half of its energy demand from, ocean currents within three miles of its shore. In fact the only attempt I know for capitalizing on ocean currents is at Florida Atlantic University in Sun City, Florida. Dr. Frederick Driscoll was building a generator designed to capture the energy of ocean currents in 2008. I suspect his project stalls because of lack of funds. Assuredly it is not for lack of potential.

Maps and charts of both shipping lanes and ocean currents already exist. The ocean current maps even show the strength of the current which is remarkably consistent year in and year out. If we superimposed a map of shipping lanes on a map of ocean currents, it would be easy to find ideal locations for an ocean current power station. It need not be large. In fact it should not be large, because we don't want to sap the energy of the ocean currents so much that they fail to accomplish their function in our environment. Many smaller ones would be more appropriate.

One advantage to ocean currents is that they will support wide variations in size. You can build a few massive ones as well as many small ones. Many smaller ones would be both easier and less disruptive. Man has a tendency to overdo what he finds effective, and ocean currents can become very effective. The tools for proper location already exist. There are some other variables that further enhance the practicality of ocean current power stations. Ocean currents are not all at the same depth. Some are deep enough to allow them to be placed in a shipping lane. Since shipping lanes are usually very close to dense population areas, deeper stations in shipping lanes may well be the best opportunity for larger stations to serve large populations.

This situation invites small enterprises to establish small power stations in an ocean current, and to build upon that base to extend it to more and larger such stations. It might be an opportunity for communities near the ocean to encourage a consortium of small entrepreneurs to build ocean current facilities to power their local community. It would be a far less tedious problem than building a wind station on public land with government restrictions and citizens complaining about cluttering up their ridge lines. In fact most communities that close to ocean currents will not have the hills on which to build wind generator facilities. Nonetheless, windmills at sea can easily be placed on ocean current facilities and capture the energy of both wind and water in one facility.

I have neglected another major advantage to developing multiple alternate energy sources. Since our demand keeps increasing, the pressure on finding new fossil fuels is immense. If we seriously cut back on our demand through conservation and alternative sources, the drive to lay waste to ANWR and other natural phenomena will decline, or even disappear. Environmentalists are frequently in conflict with major energy suppliers. Here is a way to reduce or even eliminate that conflict. If we have other energy sources, we need not drill in ANWR, or go so deep in to the Gulf of Mexico that we cannot protect and control our equipment.

The lack of effort in ocean current technology invites the use of federal stimulus funds to develop the technology to capitalize on the science that is already in place. Stimulus money has gone far too much into large corporate enterprises which make products on which big corporations make money. They employ fewer people and outsource many of their supplies from other nations with cheap labor. Here is an opportunity to direct stimulus money down to the man in the street where it will create more jobs as well as more diversity. Perhaps you can convince Uncle Sam, that a gigantic untapped source of energy is worth developing for our energy profligate populace. Give it a try!

Chapter 13

Combined Solutions

I have often mentioned that energy is everywhere, in dozens of different forms. We limit our sources of energy to two or three of these almost exclusively because of tradition—a cultural mindset that will disappear in the next quarter century. The tradition grew out of convenience as well as a lack of the science to branch out to other sources. Today that science exists, and much of the technology to capitalize on it also exists in small pockets spread sparsely all over our society. The "New" and unusual sources of energy have been there all along, but we have ignored them. We have increased our demand so much that we can no longer neglect these alternative sources. The trick is to get all of us to do it. As I said before, it is not Big Energy's job. It is ours. It will involve most of us, and all of our children. Let's take a look at some potentials.

Combinations of alternate sources are an important part of tomorrow's energy. Combinations broaden the array of solutions to the energy squeeze. I have already mentioned two kinds of solar panel for your roof. Currently they are expensive, but hopefully the price will come down substantially. What about a new version of that solar panel? It has photovoltaic panels mounted on top of a fluid base that absorbs the infrared radiation that has passed through the photovoltaic layer. Combine the two solar devices. Your objective is to get electricity out of the top layer and heat out of the lower layer, thus using a much larger part of the total solar radiation. It will remarkably reduce the cost to output ratio of the rooftop installation. For about a 20% increase in the cost of installation, you can double the output of your rooftop panels. The heat can be piped through your baseboard heating system or your hot water heater.

None of these should be the sole source of that energy. You will still buy electric power from you public utility, and will still heat some of your water or house with fossil fuels. But you have seriously reduced your expense for all of that, and you have also gained a degree of independence from energy supplier. If there is a power failure, your computers, LED lights and phones will still work. You may not be toasty warm, but you won't freeze or damage your piping system. You can still drive your hybrid car that is charged by your solar panels.

I envision an increased use of geothermal energy, but caution you that installing geothermal heating and cooling may be cost effective for a new structure where you have excavation going on anyhow, but a retrofit system may be too expensive to be economical for an existing building. Remember, you are not only trying to gain a degree of independence, but trying to get your costs down as well. Perhaps you can devise a way to get the costs of geothermal installation down to a reasonable figure.

An interesting complication to your solar or wind powered electricity is the conflict between low voltage DC and higher voltage AC. For new structures, I strongly recommend a wiring system for each. Adding DC wiring to an existing structure is not that expensive--probably less than putting in inverters, alternators, phasing equipment or other special devices for making DC add to your existing 60-cycle power supply.

Aside from expense, there is another critical consideration in this choice. The grid is operated electronically with coded controls. Unfortunately those codes have already been compromised and are available to terrorists all over the world. They could literally shut down the grid and leave us without power for a substantial period of time. If you have a true alternate source that supplements the grid instead of adding to it, you will not suffer that fate—you will still have local power.

Tomorrow's house will probably be built with solar panels and wind generators already on the roof as well as on the garage. Solar panels may not be economical in northern climes where it gets pretty cold in the winter, and the sun never gets truly high in the sky. Where they are, they should be designed to be as economical as possible. Geothermal heating may also be cost effective for new larger buildings but rarely for small or existing ones. Geothermal lends itself to larger installations because it is more expensive to dig small holes. Conservation is still the most cost effective way to reduce energy costs. When you begin to add equipment to existing structures, pay close attention to the payback factor. You are doing this to improve the cost and effectiveness of your energy, not to impress the neighbors with how environmentally conscious you are.

Not all sources of energy adapt to the private home any more than they all adapt to massive interstate power supplies. Combining different technologies in communities may well be the best answer for many alternate energy forms. I have already mentioned the new nuclear reactors that are more suited to communities instead of interstate power systems and the Grid. A new idea is the use of thermoelectric coolers for these reactors that emit massive amounts of heat. Capturing that heat and turning it into electricity could add substantially to the output of a nuclear power plant. What's more it emphasizes the "alternative" concept in energy. That output might be in low voltage DC, in addition to the high voltage AC which the nuclear reactor generates. Combinations and alternatives are meant for each other.

Small communities or private housing groups can also tap into local streams for electric power. Ocean front communities or resorts can create energy with the flooding and ebbing of tidewater. In addition, it would be much cheaper to use geothermal energy if there are groups of homes, condominiums or apartments tapping a single larger facility. Neighborhood groups can organize their members into groups to study all sorts of alternatives to oil, coal and the Grid. They could create committees for each potential alternative source and evaluate each against the others to select priorities or to reject unworkable ones. Working together is fun and informative. We grow together as we learn together.

There are many possible ways to combine different technologies to work together for a more efficient and flexible energy system. I have already mentioned the use of hydrogen as a conduit for energy. Hydrogen has its problems too, but they probably are soluble. We are misleading the public about our current use of hydrogen, most of which today is made from methane. We are fooling ourselves by doing that.

Putting both solar and wind on a single home may be more cost effective because they don't generate power on the same schedule and have substantial seasonal and geographical differences that make them supplemental to each other. I look forward to the day when all new construction has both solar panels and wind generators designed into the original structure along with a dual wiring system for both DC and AC. I am convinced that with today's technology and equipment we could build completely energy independent homes at a reasonable cost. Why don't we?

I have mentioned the use of hydrogen dissolved in other fuels. Such a plan may substantially enhance conventional fuels and make remarkable reductions in greenhouse gas emissions. They would be considerably cheaper if they are planned correctly. I see this as an opportunity for a complete new industry with almost unlimited potential. So far it has not even been given a trial, although I don't know why. Combinations of locally advantageous alternates will make surprising changes in our overall energy performance. We aren't just trying to reduce our carbon footprint; we are also reducing our overall cost and our overall risk. Neither should be done at the expense of the other. They should be planned to work together. You undoubtedly have some new ideas of your own. Give them a try, and you will learn how to make them better. What's more, you will involve your friends and neighbors and have some fun doing it. Learning new things together is probably the best basis for establishing long-lasting friendships. Have discussions at your local library, Town Hall or church. You will be amazed at how much you enjoy it.

You will also find that the Internet is a giant source of energy data and ideas. Get in touch with various colleges and universities. Almost without exception they have energy programs not only for their students, but for local people. There is no perfect source of energy. They all have some problems and some advantages. If we study those and combine ones that mitigate the problems of others, we all progress toward the new world of energy that we are creating right now. Join the fun.

Chapter 14

Do It Yourself Energy

Most of "Re-Volt!" is about your involvement in the energy revolution. There was a whole chapter on conservation, the fastest and least expensive way to reduce our energy consumption. Conservation will pay you right now! But conservation alone will not solve our energy dilemma. When it comes down to purchasing new equipment, or modifications of your current energy resources, you could always use a little guidance. Of course, as I have said before, evaluate the payback period before you make any substantial investment in energy savings. As you do, as we all do, there will be a major shift in the supply as manufacturers compete for a new and growing market. They'll try to give you a bigger bang for the buck so that you'll buy their product. Long term, I anticipate that cost of most alternate energy equipment will be reduced by about half in the next decade if we all do our homework and demand cost effective devices. I hate to say this, but over the last twenty years, most manufacturers charged quite a premium because their customers wanted to be recognized as conservationists and were willing to pay a premium. In addition, even when challenged, they grossly overestimate the output of their equipment, and quote unrealistic payback periods. That view is gradually disappearing as more ordinary citizens want to conserve energy, but do not necessarily want to brag about it to the rest of the world. Conserving energy is fast becoming a financial need as well as an environmental need. Demand to know the payback period of equipment you buy, and challenge the answers, because most are calculated on the rare, and often unachievable maximum output of these new devices.

The new view of energy is finally coming into its own. There are dozens of clever small companies making windmills solar panels and other natural energy devices, but there is something lacking in most of them. The windmills, for example are shrunken copies of the big utility windmills, but are going to be put in urban or residential areas where the wind conditions are drastically different. They won't work well there and need newer, more responsive designs. Very few of those exist. The usual way of getting the cost down is to buy used equipment at a discount or to do most of the work yourself. Not very many people are that handy. This book is trying to stimulate creative new approaches to alternate energy. No one is making the light (urban) wind generator I disclosed, with a cowling, blades like a window fan instead of like an airplane propeller that makes low voltage DC. Energy is so diffuse and so widely dispersed that it is an open invitation to your creative ideas and designs. That's why I continually stress getting involved, and joining your friends and neighbors. You don't need to be the local handyman, but you probably know one! Opportunity abounds!

Man is a prodigious consumer of energy. But, as I have pointed out often before, the sun puts more energy on this earth every day that all mankind uses. But we have a culture that uses the sun's energy from stored sources instead of directly. Even more important, however, we have a cultural bent for spreading out and using energy consuming vehicles to move us ever further in our daily lives. Transportation has been gradually wedded to oil since the invention of the automobile around 1880. Even our food and clothing are dependent upon oil for the trucks, planes and trains that deliver them. Why do we always pay some middleman to deliver the sun's energy? Why don't we go direct, or find another source of energy? The answer is simple. It is easier. Not better, mind you, but easier! For many years we didn't know how to go direct to the sun. Today we have that capability, but we still haven't built much of the equipment required to go direct. It is your opportunity to create new devices to capture directly our largest source of energy. These devices are long overdue, so the market will respond to creative ideas quickly.

At least part of that is our changed mindset. For many years we were proud of our ability to make things better than anyone else. But with the computer and the Internet, we have gotten sidetracked. We talk and plan and write about new things, but we don't make them. China does. We have given away our remarkable mechanical facility by which we built most of the new things in this world. So the manufacturing jobs we sent away were never replaced by new things to make. We have forced a change on our mechanically skilled populace so that we no longer use those talents. Energy is an opportunity to reclaim one of our most precious cultural skills. We have thousands of things to build, and we have millions of people to do it. We should retrain ourselves in those manufacturing skills which once made this country great. If we do, we will again lead the world in making things that work. We will grow our export market and pay off our national debt. We will employ our people.

Our current sources of energy are already declining. Over the next century they will become overstrained, or even used up--whether or not we like it. They already are. The Energy Revolution is upon us. It will affect everyone. The price of energy will increase as we slowly change our mind set away from the "conventional" energy supply. To protect your pocketbook, you have two choices. First is to reduce your demand for energy. Second we must create new devices that go directly for a reasonable cost. Count on it. The cost of energy will increase for at least a decade. With today's technology it won't be hard to make the change, but it won't be fast either. Your first action is to reduce your demand. It will take your personal initiative, because this change will be cultural. We must learn to think in terms of going to work without traveling 50 miles in a vehicle that weighs 30 times as much as you do. We need to shorten the distance, decrease the number of trips and shrink the vehicle or put more people in it. The Internet is one tool for the second action and car-pooling or public transportation is a tool for the third. The Internet is popular, so that part of the transition will not be hard, but public transportation has some pretty serious limitations. People are spread all over the place, so bus and train routes are an intricate tangle and a major challenge. This will require a cultural change. The house in the country is probably the largest user of energy, and particularly the transportation energy. I suspect that big companies are going to be building many small offices scattered throughout residential areas and interconnected on the Internet or by cell phone. The massive office building in the big city is an energy efficient structure, but if it requires its 3000 occupants to travel 50 miles every day to work there, it is an energy disaster. Let's face the fact that our American culture was designed for and grew up on the back of the private automobile. It will not be easy to change that scenario. It will also meet opposition from all those people who grew up dreaming of living in a giant house on three acres full of trees. Have you noticed that the first thing people do when they become multimillionaires from whatever source is to build a country estate that consumes massive amounts of energy and requires miles of travel to get anything done. The concept of the NY cliff dweller who can walk to work, to the theater, to shopping and to the subway is hardly the American dream. About 20 million Americans love it in about a dozen cities!

For decades, oil, coal and electric companies supplied our basic energy needs. With our help and support, they worked themselves into a position of dominance little short of monopoly. Some of them got carried away with the profit end of the business, and neglected the service they perform. The very fact that Exxon Mobil made the largest corporate profits ever recorded in all history and retired their president with a $400 million retirement package when Joe Workingman couldn't afford to buy gasoline was a wakeup call for the American people. That will not soon be forgotten, and while there is venom and hatred for Exxon Mobil, there is also an underlying current of determination not to let that happen again. Here is your chance America. Find, develop and exploit dozens of alternate ways to get the energy you need for everyday living. Start with the ones that are ready for the market:Solar heat, Photovoltaic panels, and wind. Geothermal and tidal energy are still in their infancy, and either very expensive, or just not available. Both the science and the technology already exist. Conservation, innovation and determination are the keys to that change. That will happen, although probably not overnight. The attitude changed overnight, and it is still growing.

Such action begs the public to replace giant businesses as controlling energy suppliers. *Re-Volt!* will help us all do just that. It will not do it <u>for</u> you, but it will help you do it. We will not eliminate Exxon Mobil from our energy supply, but we will remove them from their controlling position in energy. No more gouging the public for their corporate profit. We have learned.

Energy is diffuse. It is hard to pin down. We have many creative new ways to obtain energy, but somehow many of them get side tracked. Consider ethanol. The idea was to reduce our dependence on foreign oil and the gasoline that is made from it. Look what happened. The government subsidized the production of ethanol and we added 5% or 10% ethanol to our gasoline. We didn't do our homework. Corn already had a large market in food for people and livestock. We didn't realize that all the corn in the entire country would only supply about 8% of our gasoline needs. We also forgot that fermented ethanol produces equal amounts of greenhouse gas. So, the price of food increased as did our greenhouse gas emissions. That's a solution? That's a disaster—a good idea that went astray because people wanted to make money on it more than they wanted to solve a problem.

There are many such inconsistencies, which had good intentions, but went astray. Energy is indeed diffuse. But we can manage it, and we will in this century. This book is your guide to that change. I hope that it stimulates other ideas from you, the reader, because, like it or not, you will be involved with the Energy Revolution.

The prime message is that the energy revolution involves YOU! This is not Big Oil's job, nor is it Barack Obama's. It is yours and mine. What have you done to reduce your dependence upon foreign oil, or on coal-powered electricity? With a little planning almost every one of us could easily reduce our use of both gasoline and electricity by 10%. With a little more effort, we could reduce it by 20% and, by investing in energy conserving technology, by 30%. That all by itself would cut our dependence on foreign oil by half, and would make next year's CO_2 emissions less than last years. Ahmadinejad would be broke. Now that's worth doing!

Conservation has already been covered. It costs nothing, takes effect immediately, and saves you money today. It reduces greenhouse gas emissions right away. It requires lots of planning. Car-pooling, bicycling, walking, using the Internet, combining and shortening shopping or school trips—these are the transportation tools. The plan is the key. Turning out unused lights, or turning down your thermostat or air conditioner, caulking, weather stripping--these are some of the tools for the home. But again, the plan is what makes it all work. They will add up to substantial savings. They not only cost you nothing, they save your money as well as your planet. Picking one day a week when you work at home and send in your work over the Internet will cut 20% of your transportation costs and greenhouse gas emissions. If everyone did it, traffic jams would ease. Most passenger cars hold five people. They contain one. Fill them up. Or make some smaller ones for one or two people. If we work together, these and similar actions will cut energy demand--and its price.

Many of those actions require no capital outlay. They are just a beginning. Putting a solar panel or a wind generator on your roof would reverse 30 years of negligence in a few years. Putting on storm windows or adding insulation to your attic will also help. Trade in your SUV for a fuel-efficient sedan or a hybrid. If you want to beat the obesity fad, try bicycling to the store, or get your kids bicycles for school. When I was a kid, the school bus wouldn't pick up kids who lived within one mile of the school. I walked or bicycled to school for 13 years. I never took a school bus. I didn't feel deprived.

I have an idea for you to try. Form a "Club 280". That is where you and your neighbors make cooperative effort to reduce atmospheric CO_2 from the present 390 ppm to the 280 ppm that prevailed for a million years prior to 1900. Set targets and measure your progress against them. They will be difficult, because half of the CO_2 we have emitted ends up in the ocean and it will come back out as we decrease the atmospheric CO_2. That process will save lots of marine species like oysters and corals that can't form easily in acid oceans.

Most of the resistance to conservation is social. In the 1970's my wife Peg got a Day-Glo orange trailer for her bicycle and painted "Burn

Fat Not Gas" on it in large fluorescent green letters. She offended many neighbors when she went shopping on her bicycle. Why? Her bicycle didn't hurt them! On one occasion the 19 year-old checkout clerk at the grocery store said, "Oh, you're the one who bicycles to the store." Peg said that she was. "Oh, I'd bicycle too, if I had your figure." Peg was nearly fifty years old.

That's the message. We need massive changes in our energy sources. But we also need millions of small contributions, like the bicycle or car-pooling, or sending in some of your work on line. It costs money to replace your SUV, but it saves money to drive it less!

Many of us think we can't contribute to energy realignment, because it is too complicated. It isn't. All it takes is a little determination and a few minutes of careful thinking. You can find thousands of suggestions and even instructions on the Internet. You can learn to build a windmill, or install a solar panel. You can even find them on E-Bay or Craig's list. All of these aren't for everyone, but there is something there for each of us. When it comes to investing in energy saving devices, be sure to calculate the payback period. And don't take the manufacturers word for it---windmills produce about 1/3 as much electricity as their makers claim. These studies hopefully will stimulate your other ideas as well. Perhaps you can come up with an idea of universal appeal. It's worth the effort. In any event, take an inventory of your energy use and set goals for its reduction. If you spend a little time planning, you will be able to reduce your energy consumption by 20% and not feel one bit deprived. In fact you'll probably feel better and healthier too. And think of the money you'll save!

Cooperation is a trademark of humanity. We cooperate more than any other species. A few species that do cooperate accomplish remarkable things, like bees, ants and wolves and beavers. Put that cooperation to use. Form neighborhood energy groups that help each other find new sources of energy. Get on the Internet or cell phone, or go to the library to find new ways to capitalize on the abundance of energy that surrounds us. Unfortunately politics always enters into energy just as it does in any social change. Write and call your Congressmen early and often. Let them know that you don't want to see gasoline at $8/gallon in 2012. Tell them that you don't want to finance Al Qaeda and the Taliban. You will make some new friends and get some creative ideas about energy as well. You might even start an energy business.

Chapter 15

Alternate Energy Proposals

It is interesting to note that our energy sources have narrowed in the last 100 years, not widened. Only in the last decade have we seriously considered diverse energy sources that we practiced 100 years ago. I am concerned that not only are we narrowing our choices, but also we are becoming ever more dependent upon major energy suppliers like Big Oil and Big Volt, instead of more self-reliant and creative. Con Ed and Exxon Mobil have developed sophisticated systems of energy supply and refined them and fine–tuned them to assure that they are the easiest, and thus the most desirable energy sources. That keeps them in control.

Our future will become ever more dependent upon other sources of energy, and more particularly upon the individual and his ability to find energy sources specific to his particular need. We need more creativity and much broader choices for our future energy needs. Big Oil and Big Volt have indeed created huge and sophisticated sources of energy. They will continue to do so. But the big change will be that the sources of energy will become broader and more localized. The important factor is not to take anything except control away from these giants. They are not bad guys, even though specific instances of abuse abound. But keeping them in control is an invitation to future abuse, and the more control they exercise, the greater the opportunity for abuse. Why tempt them?

This book presents a variety of alternatives to our current energy supply. 90% of that supply is controlled by about 100 companies worldwide. As we have seen with Enron, that invites abuse. It will not go away until we become personally more energy independent. That will start in the elementary school classroom. I have constantly emphasized the desire for alternative sources for every kind of energy you acquire. Many fairly bizarre schemes may have specific applications that suit particular needs. Our children will need to be more aware of those choices and become more personally involved in deploying them. One way to make that system not only more effective, but more interesting, is to combine two or more new technologies to capitalize on each of them more effectively. I have already mentioned generating hydrogen as well as electricity from those variable resources of energy like the wind and the sun. These energy sources are huge, but not in our complete control. You can't run solar panels at night when the need is greatest. That requires that such energy be stored. But it need not be stored in batteries. I had considered what I called the "condensulator." This is insulation for your house that consists of dozens of alternate layers of insulation and metal foil that act like a giant electric condenser (Capacitor) in which you store low voltage electricity. So far I have not been able to find any insulator that will keep that charge over long periods of time. Perhaps you can. How about storing energy as hydrogen? That broadens the field of application. You are no longer limited to electricity but can heat your house, cook your meals and heat your water without using a single watt-hour of electricity. One of the problems with such a scheme is that making hydrogen by hydrolysis is inefficient. Can you use or sell the oxygen that you also make? Can we develop a market for home-made oxygen?

Has anyone thought of reviving the Stanley Steamer? That was an automobile powered by wood, coal or charcoal instead of gasoline and diesel oil. It was a relic by 1930. But wood pellets may be the key to its rebirth. It could be a viable transportation vehicle that uses renewable energy instead of imported oil. I'm not thrilled with the idea that it produces, more carbon dioxide per unit of power (BTU, or Calorie) than gasoline, but there is a whole school of people who consider such fuels as "carbon neutral". I don't buy that euphemism. Burning fossil fuels releases carbon dioxide that has been sequestered for about 2 million years. Burning wood releases carbon dioxide that has been sequestered for less than 100 years. That's an improvement? Not if we consider that we have also destroyed the tree that removes the CO_2 from the atmosphere.

The Steamer took several minutes to warm it up enough to get any power at all. Today that is no problem. Many cars have a remote starter system that fires up your automobile and its heater while you are still eating breakfast. That is particularly popular in the colder northern climate. Such a system would make the Steamer competitive on a convenience basis as well as on an environmental one. It would already be competitive on an economic basis.

One problem with the Steamer is the ash it produces. Gasoline and diesel fuels produce lots of gaseous waste, but no solid waste. The Steamer produces both. But Steamer service stations could easily establish ash facilities and exchange wood pellets for ashes from previous purchases.

Make the pellets from dried switch grass; a fast growing nuisance plant that could easily be used as a base for making wood pellets. (Switch pellets?)There already are elaborate schemes to make liquid fuels from switch grass to capitalize on that carbon neutral concept for transportation fuel. But that idea has lost sight of the purpose, which is to make a fuel that powers an engine, not just a gasoline engine. The process is costly, complex and energy consuming. Going directly to a pellet powered vehicle would be much easier and more energy efficient. That leaves the ash problem. Problem? Or opportunity? Living plants all extract needed minerals and chemicals from the soil in which they grow. These necessary minerals become the ash that is produced when they are burned for their stored energy. Why not make fertilizer out of the ash? Return that extracted mineral base to the soil to help future plants grow. The service station could make a little money selling ash to fertilizer companies, or even to its own customers.

Sawdust is a pretty common waste product, and now is somewhat of a nuisance. Mixing sawdust with switch grass may even make more energetic pellets. We could have several grades of wood pellets like the various grades of gasoline. We need to escape ossified thinking if we are to bring energy back to the people. The Steamer is only the seed of an idea. You can make it better with your ideas. Can we all be better off if we revive an old industry that served its purpose, but got out-marketed by competitors in a different era with different demands?

A steam driven or external combustion engine vehicle could be a hybrid just as easily as can an internal combustion engine vehicle. Steam should not decrease your energy options, but increase them. Wood and pellets are not the only sources of energy that can drive a steam engine. Biodiesel or ethanol could also. Do not make ethanol by fermentation. They are developing a process for making ethanol directly from water and CO_2 with algae as the catalyst.) With our remarkable new (less than 100 years) technology, we should be able to make a steam engine as energy efficient as an internal combustion engine. That would not answer all our energy problems, but it certainly should reduce our dependence on imported oil. Isn't that a major concern? Must we continue to finance Al Qaeda and the Taliban?

I have already mentioned the opportunities to make electricity in your own home. Solar panels, photovoltaic panels and wind turbines are all suited for home energy production. If we stress that they will become more so and more people will create small niche uses that serve the purpose of diversity. The theme of this book is diversity, and personal involvement. Here is another idea. I have already mentioned that hydrogen is a good energy conduit, but not an energy resource, since it does not occur naturally on earth. Nothing says that you have to use pure hydrogen. Blending hydrogen with other fuels may offer an opportunity to overcome the major barrier to hydrogen usage—its extreme light weight and low boiling point. Dissolving it in other fuels could be the small change that brings hydrogen to common usage. I still do not recommend it for transportation, but it could serve well as your home heating fuel and your hot water heat. Your stove can run in a hydrogen blend without putting in a very high-pressure storage tank. If we are thoughtful, we can figure out ways to use existing equipment for new sources of energy. If we have to create a whole new system of delivery for some of our exotic new fuels, that makes the challenge greater, but it doesn't eliminate the opportunity. Besides, many low-pressure fuel delivery systems are already commercially available. They could easily be adapted for use with hydrogen/fuel blends. Has anyone even tried?

Combining wind and ocean currents as sources of electricity or hydrogen can also bring together different sources for an improved output. I have already mentioned that ocean currents are steadier, more predictable and more energetic than wind. I also pointed out that wind is faster, steadier and closer to the earth over water. Why not put windmills on top of an ocean current turbine and use both to generate either electricity, hydrogen or both? We can double the output and effectiveness of the generating station.

There must be a hundred ways in which combinations of energy resources can improve the output and the cost-effectiveness of alternate energy. In keeping with my theme to put energy in the hands of the people instead of the massive corporation, I suggest that stimulus money be apportioned to local creative enterprises in small amounts rather than in massive donations to giant profit-driven corporations. Let's create more jobs and more diversity in energy. It should be made everywhere, not just in giant turbines at dams or huge windmills in wind corridors. We should take a page from Mohammed Yunus and his Grameen bank. He proved that thousands of small loans are a better investment in our people than massive ones to big corporations. They create four or five times as many jobs as well. What's to lose?

We are currently seriously concerned that our stimulus money is not producing jobs. Of course it isn't. We put it into big corporations that have an ossified employment structure. Put that money in small doses into the people with a task to achieve---abundant clean local energy. It will pay off many fold in energy, in jobs and ultimately in cash flow. In fact it will produce jobs and cash flow faster than the current system, which is controlled by bankers who have historically protected the bank instead of the customer. In the last thirty years or more, this country has consistently underrated the strength and ingenuity of our everyday people. We favored ever-larger corporations and bigger businesses. If we want to change our recession, we need to change the system that created it. Here's at least one method. The same could apply to the health care industry. 20% of all your health care expenses end up in insurance companies. If we cut that amount by 60% we could automatically increase the number of families covered by 12 %.

Talk about screwball ideas, try this. There is an energy source that has not even been tried called the thermoelectric generator. Most of us have used a thermocouple at one time or another to measure temperature. The principal is that if we clamp or bond together two different metals they establish an electric potential between them because each metal has a different affinity for electrons. If two wires of different metals, such as copper and iron, are attached at each end, they create a tiny electrical pressure (voltage) if one end is hot and one cold. This, called the Seeburg effect, is quite small--microvolts per degree of temperature. If the temperature difference is high, the voltage can become large enough to drive electrical devices. You probably won't be able to run your washing machine but you could run a computer or even a telephone.

Consider another possibility. If the voltage is low, you might be able to get some real energy if you get the resistance very low. Suppose you made a copper rod half an inch in diameter and 100 feet long. Attach it to a 100 foot iron rod at both ends. If you submerge one end in the ocean the lower end will be about 30 degrees colder than the upper end and there will be about a 5 millivolt charge from end to end. Resistance however will be about 0.3 milliohms. That would put about fifteen amps of current through that system. There must be a way to make use of such a current even at such very low voltages. Any ideas? You are converting heat to electricity here, so you probably will need some large plates down in the ocean to capture all the heat you are converting to electricity. Wild idea? Sure, but it might work. Got a better one?

The average home has three major energy expenses. You heat and cool your home mostly with fossil fuels, such as oil or gas. This is a considerable expense. The average home costs between $3,000 and $8,000 annually to heat and cool. You drive your car over 20,000 miles per year. At $2.80/gal, that's $2200 per year. Gasoline will not stay at $2.80/gal for very long. You buy electricity from your local electric company. Average electric costs are about $1800 per year. (If you have an electric hot water heater, you can double that figure.) Alternate energy systems should be able to reduce that total expense by half. That will save you about $3500/year. That is well worth the effort. Beware! Many alternate energy systems are grossly overpriced. I recently looked at a solar hot water system. It will cost over $10,000. It will take over six years to pay for itself, and if I finance it, almost ten years.

Chapter 16

Opportunity

Look at the bright side of the Energy Revolution. Developing our many energy resources will create jobs for everyone. It will, if we do it right, capitalize on the creativity of our people. We will need to build hardware to make use of all those scientific studies. We will make mistakes in the process, but that's how best we learn. Subsidizing corn for ethanol was a mistake that should teach us something.

The oil companies will not solve the energy crunch. The government will not solve it. If it is to be solved, *you* will solve it. That is what we want, isn't it? The Industrial Revolution was the movement of people from the farm to the factory. They left the farm not because they were forced to, but because the factory provided a better opportunity. The Internet revolution occurred because people wanted better communication methods, not because they were forced into it by unemployment. The Energy Revolution should be the same. There is immense opportunity in moving control of energy from a few massive corporations to scattered small sources run by ordinary people using ubiquitous bountiful sources of clean energy. With our present technology we can eliminate the middlemen--the oils and the coal--and go direct. We started to in the 1970's after the first OPEC oil embargo. Had we continued our efforts we would now be well on our way to clean energy independence. We abandoned that opportunity because oil was easier. Instead, we moved from 45% of our oil by import to 70%. We now have to play catch-up for thirty-five lost years. We don't want to eliminate oil or coal, but we want to sharply reduce our usage and our imports or more realistically, our dependence. I propose that we set targets. Cut our use of oil and coal by 50% by 2020. Cut our oil imports to only 30% by 2015.

This is somewhat akin to the illegal immigrant situation. Illegal immigrants, unlike legal immigrants, come here not only because we have opportunity here, but primarily because they are forced to by lack of opportunity at home. If we do it right, The Energy Revolution will not be an escape from a shortage so much as an opportunity for improvement. We all will benefit from it. Don't wait until your back is to the wall. We are close already, but if we act now we can create jobs and empower our entire economy. In the next 50 years we will add about 60% to the world's largest industry. That will employ millions of people, and they can begin today. None of that increase should be from imported oil, which should decrease by two thirds.

There is an additional incentive. When I learned to drive, gasoline cost $0. 18/gal. When my granddaughter learned to drive it was $4.50/gal. That's 25 times as much in two generations. For a society that is built on cheap transportation that is a disaster. If we don't take action, it will get worse. The dynamics of world energy are dominated by egotistical politicians who want to become world leaders. Shall we let them do so at our expense?

Will you be a lead player? To find out, answer this question: "What have you done to help your energy crunch in the last week? --In the last year? "If you're like most of us, your answer is nothing, except perhaps complain a little. That's because most people don't know how. But they can learn. The problem is everyone's. The solution will be everyone's. Get involved. It will not only help your future it will be challenging and fun. You will meet clever and motivated people. You will learn fascinating new things. You will surprise yourself with your own creativity. *Re-Volt!* is your guide to that action. It is a compendium of new energy ideas. Even more it is a stimulus for your energy ideas. It is your guidebook for the movement of energy sources from the big corporations to the people. Join in.—You'll love it!

Take a critical look at your home, as a starting point. You use electricity from a public utility. You buy gasoline for your car and yard equipment. You use oil or gas to heat your house. Depending upon where you live, that mix can vary substantially. Look at each source and ask yourself---can I find an alternate source? For electricity, you might consider a windmill or a photovoltaic solar panel on your roof--or both. For gas you might seek a local farm that captures methane from cow manure. Methane is an even worse greenhouse gas than carbon dioxide, so burning it *reduces* pollution. It supplies more energy per pound than propane or gasoline. Remember, one of the objectives for alternate energy is to cut back on our imports, or cut them out altogether. Could you use a watermill in your local stream? If you are building a new home, consider geothermal heating (and cooling). Options abound, all you have to do is to evaluate your options and find a way to use the best of them. You can find them on the Internet, and can even find the equipment to capitalize on these other energy sources on E-Bay or Craig's List.

The problem is a challenge to ingenuity—your ingenuity. Don't rely on Big Energy or the government to solve your problem. Think up new ways to get the energy your life requires. Talk it up with your neighbors, and fellow townspeople. You will not only meet new neighbors, but you will learn a lot, and have a ball doing it.

This is no casual matter. We spent over 300 billion dollars in Iraq. Had we spent that money on our schools and on anti-drug (drug education) programs, we would not only reduce our devastating drug usage, but we could bankrupt Al Qaeda and the Taliban. What's to lose? There are dozens of organizations trying to help us change the energy environment. Most of them have good ideas and even better intentions, but they won't all apply to your situation. Look into them and find the ones that suit your circumstances. It won't take more than a few hours on the Internet or telephone to make a plan for your own personal contribution to alternate energy, to reduction of global warming and to saving money. You won't find a better use of your time.

Consider wind, water, solar, geothermal, tidal and even wave energy. None of these is a complete solution, but each of them has a contribution to make. Combining two or three may make the perfect solution for your particular problems. Start with a study of your present energy supply. Select specific areas where you think a change will make a big impact. Then study the options available in those areas. Finally, study the current technology available to put them, to use. Then take another look. Is there a way you can do better? Will a new windmill design make it more productive? Can I make a solar panel absorb more of the sun's rays? Am I close enough to a stream to put a water wheel in it and make electricity? Asking yourself these questions will be fun and will stimulate your creativity. Don't do it alone! Get your friends and neighbors and townspeople involved. Write to your senator and congressman and find if there are energy subsidies to help you get started. Find information on the Internet or at your library. But do something! That is the source of your solution. Doing it with others will make it even better.

Chapter 17

Matching The Source To The Need

Adding to or changing the sources of your various energy needs requires planning. Some sources are better suited for massive installations while others work better in small units. I have already pointed out that windmills can be in several different designs. For the utility companies that feed massive amounts of electricity into the grid, 200 foot windmills over the ocean or large lakes work just fine. But those designs won't do the job in urban areas where winds are lighter in more confined space. Windmills work there, but they are of a completely different size and design for very different wind conditions as well as different demands. The problem is that manufacturers of small wind generators accept the designs of the large utilities that had the funds and researchers to design windmills for their own purposes. Instead of being creative they tried to capitalize on other people's designs.

There are some energy sources that are particularly adaptable to large installations. The oceans almost always have waves. Most of those are fairly large and powerful. As they near the shore they become closer together and tend to peak higher and higher as the depth beneath them decreases. Those waves are massive and powerful, and require large rugged equipment to capture their energy. Making use of those waves to produce energy is not particularly cost effective for your vacation home on the beach. They could be remarkably effective for communities clustered on the shore. There are some large wave energy devices in development right now. Aquamarine Power in Edinburgh, Scotland is running performance tests on their "Oyster" which captures the energy of large waves and makes electricity. Here is one example where the big corporation has an advantage over the small entrepreneur. Aquamarine also is developing a smaller device called the Edon which may be more effective on lakes and ponds.

I have already pointed out that our ethanol craze is far more polluting than gasoline. However, there is a process which makes ethanol directly from atmospheric CO_2 using Algae and sunlight. Algenol Biofuels in Bonita Springs Florida is running performance tests on its algae process for making ethanol. The fact that it consumes CO_2 instead of producing it makes this a particularly attractive idea for replacing corn as the source of bio-ethanol for gasoline. It also is adaptable to small installations. You could put one in your back yard or on your roof. But beware! Exxon Mobil is working on an algae process to make oil for gasoline or fuel oil. They want to stay in control.

In England they are learning how to capture the energy of tides. The only commercial tidal energy station was built in 1966 on the Rance River in France. But in Bristol, England, Marine Current Turbines, Ltd has designed and tested a water wheel that generates electricity from tidal flows. They are installing a massive tidal generator off the Orkney Islands north of Scotland. Here again, they are building massive units making megawatts of electricity for use by the public utilities. No effort is being made to make small ones for the private beach home or shoreline communities. Big Energy has convinced the public that these are far too complex for ordinary citizens. That keeps them in control of energy. They see even better than you do that alternate energy is coming, so they are preparing to dominate that industry just as they have dominated the current energy system. Small and efficient water driven generators are not only commercially available, you can find them on E-bay or Craig's List. You can find small windmills there too, but most are not designed for the light winds available on roof tops. The best thing about these small water driven generators is that they can also be used in small streams and rivers. If you live within about 300 yards of a stream or river, you could get some substantial energy from those streams with a relatively small investment.

Honda motor company has also introduced a small natural gas

powered electric generator that uses its exhaust gases to heat your hot water or baseboard home-heating system. This dual purpose system is an example of the creative way to make better use of the energy that abounds on earth. I have already proposed the idea of using thermoelectric coolers in the exhaust towers of large (or small) nuclear power, plants to put that enormous excess heat to useful purposes instead of overheating streams and rivers and destroying the wildlife there.

The purpose here is to use your individual creativity to capture energy from your surroundings. Big Energy is determined to convince you that energy is their realm and you should let them do it. That keeps them in control and sets us all up for another Enron. It is critical that you gain control of your energy sources, because if you don't, prices will continue to go up, and you will have your energy sources chosen for you instead of by you. Note that Big Energy tells you how much it will cost to change our energy sources, but makes no mention of how much it will cost if we don't.

I have been critical of the hydrogen industry for aiming at transportation. But there is another possible route. Several industries have fleets of propane powered service trucks. This presents an opportunity. Propane is already delivered in pressurized tanks at about 50 psi pressure. Why not pump those same tanks up to 150 psi with hydrogen? That will increase their energy content substantially, and reduce their greenhouse gas emissions. You could run your automobile on such a system, and generate the hydrogen right on your own garage roof with either windmills or photovoltaic panels. "Dual Fuel" could be the next suburban fad. "Don't pollute when you commute!" You might even consider using wood ether (dimethyl ether) instead of propane. It is commercially available and does not come from crude oil. It very likely would hold more hydrogen at the same pressure.

Alternate energy and particularly small local sources for alternate energy is an open invitation to creative ideas. You might like to hold a local contest for the best alternate energy idea that applies to your town. Perhaps you would like to build an alternate energy house in your neighborhood--one that uses no commercial energy. It would be an educational tourist attraction.

Chapter 18

Wind and Water—Practical Sources of Energy?

Oil and coal dominate our energy sources because they are easy to use and because we have already developed the technology to use them. They have been cost effective. But they are both finite resources that are being used up. Admittedly, there is so much coal, that it will be a long time before it is depleted. Major energy companies have controlled them for 100 years. We are beginning to see the result of that as energy prices (and their profits) have skyrocketed in recent years. As oil supplies become depleted in much of the world prices will continue to rise. Many known oil resources now reside in politically unstable countries. Hence oil is a major political tool that is being wielded by despots and megalomaniacs. When any resource is so widely depended upon, it is an invitation to abuse by ambitious dictators. I note again that as the sources of oil become more strained, Big Oil has increasing profits. Is that their social responsibility? These facts make replacement of oil as our primary energy source a high priority goal.

On the other hand, wind, unlike oil and coal, is everywhere and it is not difficult to capture and utilize. It is, however, very difficult to control. Wind is not without its problems. It is intermittent at best, and usually is much diminished near the ground. In addition, most wind development in recent years has been in ever-larger units built with government assistance and subsidies. These grossly inefficient systems neglect the large majority of the available energy of the wind they capture or fail to capture. They also deplete substantial amounts of their energy in the high tension wires that send electricity throughout our world.

Water is an even bigger potential energy source. Like wind, it is dispersed all around the globe. Unlike the wind, it moves consistently and in known patterns that are already mapped. Why is it even less used than wind? We have been building dams and tapping waterfalls for two centuries while neglecting an even larger resource in ocean currents and tides. There is also a tendency to build massive Hoover Dams instead of small local units that serve local populations scattered all over this country. There has been very little local entrepreneurship in either wind or water. But that is where the majority of that energy is located.

As mentioned above water has some distinct advantages over wind. There aren't many streams that dry up altogether annually, and none that turn on and off several times every day like the wind. None of them changes direction. Water wheels and water turbines are a much more reliable and steady source of energy than wind. You don't have to build a 150-foot tower to run a watermill. In many instances you don't even have to build a dam. You can anchor your watermill in a stream or river and generate electricity 24 hours a day. But, because of our single-minded focus on oil and coal, I know of no one who does that. Why not?

An example appears in Searsburg VT. In 1997 we built eleven 170 ft. wind generators that generate 6 Megawatts of power at full capacity. The problem is that they are at full capacity less than 15% of the time, and are not working at all 30% of the time.

But look at this. Seventy years earlier they built a water-powered generator in the same town about two miles away. It is about twenty feet square and twenty feet tall, and produces 5 Megawatts of electricity at full capacity. It runs at full capacity about 45% of the time, and can operate 100% of the time. (They do shut it down periodically, for obscure reasons.) It has been working for 80 years. This is a much more effective system, and far outperforms the much more expensive windmills. I doubt that the windmills will last 80 years. In fact lightning struck one of the windmills in 2008. It had to be repaired, and blew down six months later. Even in inflation-adjusted dollars I suspect that they cost many times what the water mill cost. I can't find the facts to back this up. To be fair, there is an eight-foot diameter pipe running about three miles from a reservoir to the power plant. That, however is a local peculiarity, and need not be required elsewhere. They could have put the generator right in the dam. The earthen dam is about 50 feet high and about 500 feet long—the largest earthen dam in Vermont.

Wind is a good source of energy, but it is neither universally available nor always practical. It is much better over water where it is steadier, closer to the surface and tends to veer off course much less. You don't have to buy the land, or get special legislative approval to erect your windmills at sea. The towers need not be as tall, or as structurally strong, or as expensive. Neighbors don't complain.

Watermills aren't very practical in a desert climate. Not everyone is close enough to a stream to make use of water. For those who are, water is a more reliable and productive energy source than wind. The surprising thing is that there are almost no manufacturers of small water generators. Both watermill and windmill industries tend to go big. They play into the hands of major grid operators, so that control of energy remains in the hands of the few. Rather than putting another Hoover Dam on the Colorado River, we should install 50,000 small 20 KW watermills in rivers and streams all over the country. The majority won't need a dam. The same is true of small windmills to be located on our suburban homes. They should have plastic wind foils that capture wind velocities of 2 or 3 mph, rather than highly engineered propellers that won't operate below 8 mph winds. They should make enough electricity to pay for themselves in 3 years or less. No one makes those right now. I regret to say it, but the market for small wind and watermills is dominated by people whose objective is not to serve the electrical needs of our people, but to make money. They can do better by adding to the electric supply of the big utilities, rather than to supplement it. So far, their efforts have further entrenched the big utilities when the dispersion of energy supply is what is needed. We will not have a stable and controlled energy supply as long as it remains in the control of a few giant companies.

Perhaps you can do that job. Why don't you manufacture small windmills for the family home? Open up the small watermill market that has been largely ignored for over a century. You can find designs and instructions on the Internet. I have plans for a windmill that will generate half of your electric needs. Others have similar plans. Here is an opportunity for you to get into a new business and employ people who want to make this world a better place. If they do, they will make all the money they need. Caution: don't lose sight of the objective. It's not to damage the big utilities. It is to wrest control of energy from them. They will, and should remain a large part of our energy supply.

In summary, how should we guide our use of wind and water for energy? The critical factors and their reasons are:

- § Make electricity locally. Don't send it long distances to areas that have their own sources.
- § Make low voltage electricity so it can be produced all the time by water, and most of the time by wind.
- § Make low cost, replaceable, mass-produced units rather than high cost engineering marvels. Keep payback periods short (under three years).
- § Always assure that there are alternatives to any energy source.
- § Multiple small units are strongly favored over large ones.
- § Involve as many people as possible. Start local groups, forums and consortiums.

Chapter 19

Energy Alternatives

Throughout *Re-Volt!* I have often stressed the idea of alternatives. Man's enormous increase in population along with his massive increase in energy demand per person has overwhelmed what was once a moderate industry in a growing society. As a result, the future will demand that every source of energy will have an alternative. There are four purposes for this. First, of course, is that if a system goes down for any reason, there will be an alternative so you won't be paralyzed. Second, and more important, the long-term purpose is to prevent any company, government or group from gaining such a dominant position in any kind of energy that they can control it. Third, we will not be able to keep up with demand if we don't use all the resources we have. Control belongs with the user. Fourth, our current direction in energy usage is creating an environmental pollution problem that will literally destroy us if we keep it up. We must gain control of our greenhouse gas emissions.

This concept plays out in many ways. Here are a few examples designed to stimulate your ideas and many more:

Consider electricity. It may be the largest single source of the energy used in today's society. In any event it is a huge one. Every house is wired for 240/120 volt AC electricity. That already tells you that the power companies are in control. 120 volt AC is their stock in trade, and every one of us uses it. I have proposed early and often that we also have 6, 12 or 24 volt DC available in every home. That might require a second set of wires to carry that electric supply. It is not that big a project to install several wires in various parts of your home to carry DC instead of AC. (I understand that there are efforts to make a single system that will carry both AC and DC.) It would be an easy task to put both in any new construction. As I mentioned, the primary reason for AC is to send electricity long distances. The low voltage DC is for local use. Such wiring already exists with fittings and plugs and switches commercially available on the open market. Much of that is in the motor home and boating market because that, the automobile, the telephone and the computer are where the bulk of the low voltage DC is used today. That system already has lights and a variety of appliances on the market. It would not be difficult to assess which if any of these is suitable for your personal situation.

Using commercially available parts and equipment, your present home could be equipped for dual electricity for about $200. Because low voltage means higher amperage, the wire sizes should be about two sizes larger for DC. (There is some confusion here. Convention sizes wires in even numbers—#8, #12, #18. A size #8 wire is bigger than a #10 wire. If your AC appliance requires a #16 wire, use a #12 wire for DC, if it requires a #12 wire; use a #10 wire for DC.) I recommend that most DC wiring be #10 or #8 if they serve more than one appliance. They should not be longer than about 250 feet. Your electrician has those answers.

But having a DC wiring system is worthless unless you have a DC source. There are several. Primary are wind generators and photovoltaic panels. In most areas you might have both. Much less common and much less accessible are water driven turbines in local streams or in tidal basins. These are future sources that have not been widely produced in the past, but which will assuredly be used in the future. Note again, the purpose here is not to replace the grid, but to supplement it. I envision a day when about half of your electricity comes from local generation and the other half from the grid. Stress the "alternative" concept.

So far windmills are used primarily by the public utilities to generate electricity for the grid. That has seriously limited our use of an energy source that is everywhere. I have already pointed out that windmills generating power for the grid have limitations that make them idle more than they are productive. A good proportion of that limitation is because they need to generate higher voltage AC current. The DC system for your house has none of those restrictions. Eliminating them can increase the output of a local wind generator by from three to five fold, or more. That will greatly increase the return on investment, and cut the "payback" time greatly. It suddenly makes alternate local electricity competitive with our electric grid. The objective is to make it less expensive. Photovoltaic electricity from your solar panels can also feed into that system without specialized equipment. Many of your newer household appliances, including your computer, your cell phone and your LED lights already run on low voltage DC. Further, to increase the output and lower the payback period of your solar panels, there are new coatings for them that allow them to covert a larger portion of the solar spectrum into electricity. Their material cost is less than existing panels. There is also a system that converts the unused portion of the solar spectrum to heat. It further capitalizes on the idea of having alternates for all the energy you purchase.

If you are building a new house, consider geothermal energy. Not very far down in the earth the temperature is about 60°. That is true everywhere. If you dig down and put pipes in there, you can use that heat to warm your house--in fact it can also be used to cool it in the summer heat. But putting in such a system for an existing home might be quite expensive (at least $25,000) and probably will take over ten years to repay its cost in heat output. But for a new home where excavation is going on anyhow, geothermal may be a good alternative. Aside from the excavation, costs are about the same as current systems. They will last for many years. It is worthy of note that small installations are generally not cost effective in geothermal systems so single homes may not find it worthwhile.

There are myriad other possibilities, including solar panels to heat water, or parabolic solar reflectors which heat oil for making steam to operate equipment. Energy is present in so many places in so many forms that opportunities for creative thinking occur every day. What are your ideas? If they work, you might even consider mass-producing them and starting a business.

The most critical requirement for a future of energy self-sufficiency is the act of getting off your duff and doing something. Early in the book, I spoke of your internal energy that gets you up in the morning and lets you accomplish things. The hard part is getting your internal energy activated to accomplish something different and new to your experience. These are only a few good ideas. Clearly there are still more that don't even show up in this book. Talk is easy. Doing something is harder, but not as difficult as it may seem. Start with conservation—it doesn't cost anything and can save you some money. Take a simple example. Transportation and gasoline are major energy consumers. Probably about 25% of your energy expenses revolve around transportation, unless you live in a major city where they may be less than 10%. One very simple plan that can go into effect immediately is to decide that you will not use the car at least one day a week (or two days?). Plan to car pool with a neighbor or have joint shopping days with them. Have the kids walk to school, or establish carpools for several families. Send your work by e-mail or the Internet instead of going to work. There are dozens of ways in which, with a little planning, you can seriously cut back on your driving. Walk short distances, or use public transportation if it is available. Bicycles will also help you fight the national trend toward obesity. I cannot emphasize walking or bicycling enough. We are an obese nation, and we need the exercise. Walking and bicycling are excellent exercise and will make us a healthier nation. I would not recommend skate boards for everyone, but roller skates could be very effective for city dwellers. In the 1970's, someone introduced jogging to a sedentary public, and it became a wonderful fad. It has since declined, but I still see joggers every day. In Amsterdam in the 1970's I was impressed by the enormous number of bicycles crowding the city streets of a healthy and productive nation. These alternatives are available to almost everyone, and others are available for different circumstances. When gasoline goes to $8/gallon, these will look a lot more attractive. Make no mistake, energy cost will be a major concern in your future whether you like it or not. They will go up, not down, because demand is still exploding. Planning better usage is not that hard and is remarkably cost effective. Making your own is still better.

Consider this ploy. Require your children to exercise an hour for every hour they watch television. You might even try it yourself!

Conservation does not apply only to transportation. Today electricity is not a major part of your energy expenses-probably about 15%. That is already beginning to change, although as energy costs increase electricity will most likely increase at a slightly slower rate. A little planning will probably allow you to cut your electricity use by at least 10%, and perhaps 20%, without giving up any daily comforts. Electricity has been so available and so inexpensive that we as a culture have become seriously profligate in its usage. Even as we move toward low voltage devices, we have introduced the small pilot light on every electric device we use. Almost every home has at least a dozen items with a little pilot light bulb even when it is turned off. That will change over time, and careful planning of electric usage will doubtless become much more common. The reason for the planning is to assure that the decrease does not cause disruption of daily activities. I have already suggested that you make a plan that reduces your monthly electric bill by 10%. You will be surprised at how easy that is to do. But remember, electricity is designed to enhance your lifestyle, not limit it. Planning will allow that.

You don't have to buy complex thermal control systems to cut your home heating bill. Cutting off heat in unused rooms is not very difficult unless you have small children. They learn early how to turn up the heat, and often do. A little education will help control that. Closing the damper on an unused fireplace will save a surprising amount of heat. Having your furnace cleaned will often save 2 or 3% of your heating bill. Caulking and taping the cracks and seams in your house is surprisingly effective in cutting heat loss.

Heating and cooling your house is, in most U.S. climates, the largest single use of energy. It costs about twice as much per degree to cool your house as to heat it. Simple plans like cracking open the lowest windows on the shaded side of the house can often eliminate the need for air conditioning altogether.

These ideas have been around for decades. Few of us use them. There are many more so the important thing is for you to assess those that are available to you and put them into action. Taking the first step is the hard part. Once you have done that it sort of grows within you. Starting to do it in your own home seems to bring these myriad opportunities into focus and you might even get carried away and completely redo your way of living. You might want to invest some of your energy savings into additional energy saving systems--storm windows, a more fuel-efficient furnace, a hybrid car, a rooftop wind generator or solar panel. If you do, be sure to calculate the payback period. Saving energy is not a showpiece, but a practical way to save money while saving your environment. I have seen new building plans that cut out 80% of the heat losses in new homes for almost no increased construction costs. It's the plan that makes the difference. Want to join in the fun?

I would be remiss if I didn't mention food. Food is the one industry that is bigger than energy. Fortunately for us all it is extremely diverse already, so no major corporations control the food industry. Of course, large corporations do control segments of the food industry. Raising chickens, beef and pork are largely controlled by a few large corporations or alliances between a few large ones. In general food is a fairly free market. It is interesting to note that the food we eat doesn't even enter into the energy costs we calculate. It should. Raising beef is enormously energy demanding. Chickens are less so but pork is highly energy demanding. When you buy food, note that meat is generally two to three times as energy demanding as vegetables. These are no small contributions to our overall energy consumption. A pound of beef has used as much energy in its tour from calf to meat market as driving your car a mile. That's a lot of energy. You may not want to change your lifestyle, but you should at least know that the energy to raise meat is a significant drain on our energy resources. Stressing vegetables in your diet will not only reduce your carbon footprint, it probably is healthier for you as well. In any event, it is too big to ignore altogether.

Look at it this way. Each of us requires about 2000 kilocalories in food energy every day. To generate that energy, it metabolizes 417 grams of food each day and spews out 817 grams of CO_2 every day-300 Kilograms each year. Since there are 7 billion of us that is 2.25 billion metric tons of CO_2 every year just from our breathing. Since our world's atmosphere contains only 1400 billion tons of CO_2 that, by itself, is a significant factor in global warming. That tells us that population control is a major concern for global warming. Would that breathing was the only source of Man's greenhouse gas output.

Chapter 20

Power House

We have reviewed many possible ways to change and improve our energy supply. There is going to be a major change in our energy supply in the next two decades. If we plan, that change will be to our advantage. If we don't, we will all be victims. I have stressed the need to have alternates for every energy source you use. Since most of the energy you use is centered around your home or your transportation, let's design a set of systems for a family dwelling and for your transportation.

Look at the basics. Your home uses energy all the time. Your home is also assaulted daily by a variety of energy sources that are ignored. Why do we buy energy from somewhere else, and neglect the energy that bathes our home every day? The objective here is to use the wasted energy engulfing your home to supplement and reduce the energy you buy. Your house is a Powerhouse.

There are many alternatives to the energy you currently buy—oil, gas and electricity:

Solar energy
Wind energy
Geothermal energy
Tidal energy
Ocean energy
Bio energy

For the private home, two stand out: Wind and Solar. Consider also water and geothermal energy. These vary enormously with the geographic location of your home, but all are available. In different areas, different sources will predominate, not only because of your geographic location, but also because of the structure and orientation of the home itself. In Arizona, photovoltaic energy will probably dominate your local alternate energy sources. Wind, water or geothermal energy will probably be less cost effective. But even in Arizona, if your house has a roof-ridge that is east/west aligned, you will have an easier opportunity to capture solar energy than if your ridge points north/south. If you have flat roof, orientation of the house may not matter much. Similarly in Vermont wind will probably be more productive than photovoltaic energy, although solar heat may be cost-effective. The surrounding trees and neighboring homes will all have considerable effect on wind generators as well as solar panels.

Your goal is energy independence--not energy isolation. You should not change from dependency upon one source to dependency on a different one, but rather to make alternates available. You will still use electricity from the grid and fuel for central heating/cooling but less of each. You will not be exclusively dependent upon any of them. Not every electrical device in your home will draw from the grid. If you run out of oil, or gas, you will still have heat. Look at your options, and match them to your specific location and use. None of us will use all of these options, but all of us will use some of them. If you do your homework, you can select the most cost effective technology for your home and save much of your energy bill for the rest of your life. If you're clever it should make your life easier as well as more energy efficient. We are not trying to save energy at the expense of comfort and enjoyment, but to enhance them.

Make a plan. Remember chapter 3? Conservation is the fastest and cheapest way to save energy. It works now, and costs nothing. So look at your largest energy demand and figure out how to reduce it. Heating and cooling are probably your greatest energy demand. What is your source of home heating? Oil, gas, electricity or coal supply about 99% of our national energy demand. You can do better! Which is cheaper in your area? Your burner probably is not clean and at its peak efficiency, although fuel companies are pretty good at making an annual check on boiler efficiency. Do you even know what your last efficiency rating was? If it is below 85%, give serious thought to a special cleaning or even a replacement. Be sure to calculate the pay back factor.

Homes in the south will find different sources are more efficient than homes in the North. That is the wonderful part. You can use those that are adapted to your specific needs, and let others do the same. Calculating the relative costs is somewhat tricky, because they are measured in different units. Propane is sold by the gallon, which weighs only 4.64 lbs and provides 84, 345 Btu. Fuel oil is also sold by the gallon, which weighs 7.22 lbs and delivers 128,870 Btu's. Coal is sold by the pound, which delivers 14,030 Btu's. Electricity is sold by the Kilowatt Hour that is equal to 3414 Btu's. A kilowatt-hour of electricity is equal to burning 0.04 (1/25th) gallons of propane. If electricity costs 12 cents per KWH, an equivalent price for propane would be \$3.00/gallon. Equivalent energy from #2 fuel oil would require 0.026 (1/40th) gallon. An equivalent price for fuel oil would be \$4.62/gal. At today's prices, which probably won't last very long, fuel oil is a substantially better buy if cost alone is to be considered, and electricity is the most expensive. I did not calculate coal, since coal furnaces no longer are on the market for private homes.

If you live in a cloudy area, you won't find Photovoltaic panels very cost effective but probably will find wind energy useful. NOAA has historic cloud cover percentages for most large cities in the U.S. If you already have built your home, geothermal probably won't be cost effective, but if you are building a new home, it is probably worth considering. If you live near a stream or river, a water driven generator may be very helpful, but probably will require cooperation from your neighbors. As of today, a water driven generator is difficult to find. Hopefully that will change, but you could build one yourself with the help of your local handyman.

Direct solar energy comes in several forms. The sun has a broad spectrum of light wavelengths containing differing amounts of energy. Visible light can be converted directly into electricity with photovoltaic panels. Long wavelengths (infra-red) can be used for heating water and your house with hot water baseboard heat. Both photovoltaic and solar hot water panels are readily available in many sizes and varieties. Solar hot water systems that supply up to 85% of your hot water demand are already on the market. Some are high-priced showpieces that may take a dozen years or more to pay for themselves, but recently there has been a substantial increase in the practical panels that focus on saving money. Both the heating and the electric panels are also available in do it yourself kits. The Internet also has instructions that claim to put solar heating in your grasp for under $200. All of these are available on the Internet. Here are a few sites, but far from a comprehensive list. Look for solar panels on the Internet:

Sunbug	Solar Solarpanelinfo	Solar Home
GoSunward	Solar Depot	Siemens solar
ServiceMagic	Get Solar	

This small sampling of outlets runs the gamut of low to high priced and from professional installation to do-it-yourself systems using used parts from E-Bay or Craig's List. Some supply Photovoltaic panels for electricity, and some supply solar heating panels for heat and hot water. Some supply both, but none that I have found supply a stacked system that uses a photovoltaic mounted on a solar heating panel to capture essentially all of the sun's energy. However, there are parabolic reflectors that concentrate all the sun's rays onto pipes containing oil that get heated to 700°F. The hot oil can boil water, or drive machinery to put all that energy to good use. These are usually fairly massive installations and used industrially rather than in private homes. That may change. Perhaps these are opportunities for you to start a new business.

Wind is the second system that is particularly adaptable to the home. As I have been reporting in previous chapters, the windmills you see are ones of utility companies, but not cost-effective in the private home. Like the solar, which has different mechanisms for capturing energy from long and short wave light, wind has at least two separate systems—Giant heavy-wind propellers for industrial use and small light-wind generators for private homes and small installations. It is my hope and belief that the future will see a complete development of the small wind generator that should be on most of the homes of the future. Current supply of light wind generators is severely limited, and has been confined mostly to the do-it-yourself market. The ones that are commercially available are mostly of obsolete design and are not cost competitive.

Geothermal energy is more of a challenge. It is universally available, but requires some pretty costly preparation to utilize. Digging holes in the ground is not for the man with a shovel. Yet the ground is where the geothermal energy exists. It also requires special refrigerants to make use of the relatively low temperatures available. I hope that some clever individuals reading this book will find a solution for the high cost of installing geothermal energy. Once the system is installed, it can operate trouble free for long periods of time.

Chapter 21

Transportation

Transportation consumes such a large portion of our energy that it deserves a separate chapter. Today transportation is remarkably limited and outdated. Almost everything we do requires an automobile. Not only an automobile, but the gasoline powered automobile. Our railroads are falling apart. We don't have high speed rail. The train is no longer a competitor to the private automobile. When we began building superhighways beside railroad tracks, the rail industry gave up on moving people and converted to a freight system. It has become a ghost of its former self. Street cars are gone. Even buses are a shadow of their former glory. Bicycles have almost disappeared, although there has been a small re-emergence of them in the past decade. Walking is a forgotten art, although jogging for exercise is experiencing a comeback. This is not focus—it is obsession!

Why do we so limit our transportation? The American automobile is a powerful gasoline driven status symbol which helps define our personality. Our view is so limited that we neglect not only other modes of transportation but even most variations on the car itself. There are few diesels, no steamers, no electric cars, no propane cars. We have established a national, perhaps world-wide, mindset about transportation. The automobile has even replaced walking. We are becoming a nation for whom things are done, rather than one that does things. No wonder we are getting obese.

In 1900 people worked at home, or walked to work. As early as

1827, horse-drawn public transportation was available in a few large cities, but less than 1% of people used them. In 1863, London introduced the first underground railway system. It was powered by steam locomotives, which were smoky and stinky. It converted to an electric system in 1890, the world's first electric powered public transportation. In 1871 San Francisco introduced the steam-powered "Cable Car" which replaced horse-drawn streetcars.

None of these could keep up with the rapid suburbanization of the United States. Under heavy pressure from Henry Ford, our country moved overwhelmingly to the private car, and to ever more distant suburban homes. The net result is that transportation is the country's largest single user of energy today.

Transportation becomes ever more challenging, and ever more stultifying. I suspect that nearly half of all the families in this country would not survive a month without a car. Consequently it is worth our while to study transportation and see if we can make some major improvements in it. The biggest event in a person's life today is the 16th birthday and the driver's license. We hardly even notice the eighteenth birthday and the ability to vote.

The automotive industry is massive, profitable and entrenched. It is resistant to change and it exercises control of any change. It will be difficult to alter the current trends in transportation because the automobile manufacturers protect and exercise their control. The problem is that a change is required. The longer we delay the more necessary it becomes—and the harder it becomes. We have a dilemma which is also an opportunity. Individuals will not be able to move this giant industry, but groups of creative people can. It will take lots of effort and will face direct confrontation from a massive establishment. That is the challenge.

The first automobiles were primarily curiosities. They were big and loud and tended to be pretty fancy. Before 1910, many of them were electric cars. These had limits of about 40 miles on a charged battery, and could rarely go over 20 miles per hour. That was adequate for the circumstances of the day-- few roads, short distances and no traffic. The electric car couldn't keep up with a rapidly changing society. The distances demanded of an automobile for transportation grew from about 20 miles to several hundred. The growth of the automobile hindered the growth of our railroad and our public transportation systems. Only in about thirty major U.S. cities is public transportation a major contributor to daily life. We have no high-speed rail, although Europe and Asia have several such systems. Our patterns of living changed drastically to make use of the automobile--and to become dependent upon it.

But things are still changing. The overwhelming use of cars today is commuting to work. Over 80% of the work-to-home commute is in cars with only one person inside. We have created a system in which waste is the major factor. Why should you drive 20 miles each way to and from work in a car weighing over 3000 lbs? There are electric cars today that will go up to 200 miles on a single charge and can go over 50 miles per hour. There are three major commercial electric cars in the U.S. today--the Chevy Volt, the Nissan Leaf, and the Tesla. Many people don't even know that they exist, so there aren't very many. The electric car has a technological handicap. It is not possible to store as much energy in a car battery as there is in 12 gallons of gasoline. In addition, today's batteries take three hours to charge. You can fill your gas tank in 5 minutes. That is a reason that there aren't any roadside recharging stations, and there probably won't be.

Let us find other sources of transportation energy: Wood pellets for an external combustion engine, blended fuels, vehicles with solar electric power or from windmills, dual fuel vehicles, mopeds, bicycles, motorcycles, trikes. The list is endless. Multi passenger vehicles like buses, trains, streetcars, trams need to be re-developed to meet today's demands. There are hosts of alternatives many of which may do the task at hand even better than the automobile. Transportation is far too encompassing to be limited to a gasoline powered private vehicle. That limitation is culturally derived, not functionally derived. We select transportation because of societal factors rather than functional factors. That must change. It will be more of a challenge to our culture than to the individuals within it. The American car is far more of a status symbol than a transportation device. That should change.

There is a ray of hope. The hybrid is the only significant automotive change in fifty years. It is an entry point to a whole new system of transportation --dual energy sources. Today the Hybrid blends electricity with gasoline. It is the very symbol of "alternative". That is a good start, because both systems are in advanced development. Alternate sources of energy are a key to energy independence and energy freedom. There is no shortage of such options for transportation, but they are almost universally ignored. Many have been used in the past. The electric car was killed by our moving farther and farther from our workplace. We outdistanced the electric car. Solar and wind recharging may be able to revive it. The opportunity to revive the electric car is growing as we also distribute workplaces and shopping centers ever more widely into the suburbs and the country. Commuting distances show signs of decreasing. Better storage technology for electricity has increased the range of an electric vehicle. But today, if you have an electric vehicle, you must charge it at home. There are no electric stations. There probably won't be, because recharging a battery takes about 50 times as long as the refueling of a gasoline powered car. Could we establish exchange stations where you replaced your spent battery with a recharged one? Electric transportation is better for dual sources of power. Gasoline, diesel, propane, wood pellets—each could find a market for hybrid electric vehicles. Recharging systems also have many options. Photovoltaic panels on the garage will not likely be much use, since the car is out when the sun shines and at home in the dark. But windmills or water wheels in your local stream show much promise.

Many years ago, my late wife and I spent our honeymoon on Bermuda. We rented two mopeds and there wasn't a square inch of Bermuda we didn't see on those wonderful vehicles. They were fairly sturdy bicycles with a little (2 hp?) gasoline engine at the rear wheel. We pedaled more than we used the motor, but admittedly, Bermuda is pretty flat. Remembering that, I recently looked up Mopeds on the Internet. There aren't any! (I see that Bermuda still has mopeds, now called "Pedal Bikes". They are much fancier than the ones I remember, and appear to be more powerful.)There are 30 HP and 50 HP scooters that they call mopeds. They don't even have pedals. You can buy pedals as an accessory, but it is clear that the world has changed. Why pedal when the motor will do it for you? Pedaling might even help you keep (or perhaps regain) your slim youthful figure. (That, too, is a relic of the past, as a visit to your local middle school will show you.) I would like to see some real Mopeds used on our gradually reappearing bicycle paths along city streets. I'm not talking about a 50 horsepower scooter that will go 75 mph. A true moped would be welcome on a bicycle path. They would be ideal for suburban and small town travel. You need not buy an SUV to carry the dogs with you. They can run alongside the bicycle and guard it while you are at the PTA meeting or shopping. Perhaps a little rough in Vermont in February, but we don't all live in Vermont. Even Vermont has sunny days. Could we create a system with one set of cars for the commute and another for trips and vacations? Perhaps you could rent a car for your vacation, instead of owning it year round and paying its expenses. And don't forget not going to work at all one or two days a week and sending in your work by Internet or telephone.

When considering these alternatives, remember that the objective is to make life better, not just more energy efficient. These are new applications of old concepts, and are not currently available. Look carefully at them before starting a new venture. You have to compete with an industry that has been working on convenience for over 100 years, so the challenge is huge. The likely successes will be those that add to the existing culture rather than replacing it. The need is there and growing. If you focus on need, rather than status, success will follow. This is a challenging opportunity. It is also a source of jobs.

Step one is to initiate a second energy sources for all vehicles—from the moped to the hybrid. Electricity is clearly one second option, but not the only one. It need not apply only to the gasoline driven vehicle. It can apply to diesels, and to steamers. It can supplement wood pellets as well as gasoline. We are talking about establishing a whole new system of transportation or better still, several. They must be competitive. Unfortunately, transportation is such a basic in our society, that it must be competitive not only in transportation, but in social mores as well. That is the challenge.

We have also developed the Mall in the past 40years. You no longer have to go downtown to shop. An electric car would be perfect for such a shopping trip. In fact inner city stores have been struggling recently, and most are building satellite stores in surrounding malls. The need for the gas powered private automobile is actually decreasing, but their number and use still increases. Statistics will show that the number of trips in a private car have decreased in the past decade, even though the total miles per car increases as we still move farther and farther into the country. The continuing increase in cars per family is driven today by cultural mores rather than actual need. For your future, you should look at that carefully. Gasoline prices and pollution will continue to increase. In fact the biggest impact on gas consumption in recent memory was when gasoline peaked to over $4.50/gal in 2008. My guess is that drastic reduction in miles driven would happen again if we put a $1/gal tax on gasoline. People would scream but they would also become more realistic about their use of automobiles.

There are enough options today that the change we face should not be a hindrance. With planning, they might even be an improvement. Make no mistake, a change in transportation is coming whether we like it or not. There are electric cars, bicycles, motorbikes, public transportation, and Internet communications. The ideal would be to have dual fuel vehicles that will provide options to the owner and competition for the supplier. There are car pools and shared shopping trips. There are school buses and shared school trips. Although the change seems a little daunting, if you plan well, you might find an easier way to manage your daily life. That should be the objective. Don't make a change for change sake. Make it improve some daily or weekly task. You need not do all this at once. Take it step by carefully planned step. Begin with figuring out how to cut back on your driving. Set a goal. I now drive 200 miles a week. I will cut that to 175 miles per week. When I accomplish that, I will aim for 125 miles per week. Join your neighbors and friends. Form clubs and associations. Work with your PTA or PTO to minimize school transportation. Bring it up at your sewing club, your church group, your Rotary club or your Lions club. But remember, the objective is to make your life better, not just more energy efficient. Use the energy issue to bring together the myriad new inventions that have been created since we began to shift all our transportation to the private automobile. That is not the best system today but we haven't faced up to that yet. A more energy efficient society *should* be more effective and rewarding. It will be if we face the marvelous inventions of the past three decades and put them to use in our daily lives.

It is important that you become involved with the general transportation system. The automotive industry is immense, and powerful. They defend and protect their monopoly on transportation. It will take lots of effort to persuade them. You are not going to introduce a hybrid car that uses hydrogen blended fuel and electricity, but if enough of us get together and demand it, the automotive industry will-- reluctantly until they find out how to make lots of money on it.

Should you push for bicycle lanes on your local streets? I see elder citizen's buses taking old folks like me to and from shopping, the dentist and the doctor. Will your community support a local bus service? Will it really enhance transportation? Some cities have even found it worthwhile to rebuild the street car, although our urban design makes that impractical in most communities. An electric tram system on rubber tires seems more practical in most cities.

It is worth noting that the move to ever more spread out suburbia has also served to isolate us more. People have fewer casual friends and little direct contact with their immediate neighbors than we did fifty years ago. That is a loss. One purpose of the energy conservation move might be to re-establish the neighborhood. When I was in school, I knew the names of everyone who lived in every house on our street. I was with my neighbors every day. We played baseball or hockey on the empty lot on the corner. We rode our bikes up to the local drug store for Monday's half-price special "double dipper" ice cream cone. Today, surprisingly few of us even know our neighbors. They are too busy tweeting their school friends or talking to them on the cell phone. They listen to canned music and play online games. They create a circle of friends that are spread over a large area, to the exclusion of their immediate neighbors. They never <u>do</u> anything with them. It is all talk. Few of them have bicycles, and they never go fishing or ice skating with the kid next door. What a loss. Don't let your children continue that. That is where the drug dealer gets his customers. Put yourself and your kids back into community life. You'll meet some remarkable people. You might learn a lot too. You'll cut your transportation bill by 20% or 30%.

I see people on roller skates and skateboards traveling locally. I also see a few people who buy small cars. I am disappointed to see that, even after the automotive bailout, American car manufacturers still advertise the fancy car, rather than the affordable car. The Detroit mindset is still "bigger" and "more profitable". But we are in a recession, and many people can't afford the fancy wheels Detroit advertises. In addition, as I mentioned in an earlier chapter, the automobile is still just arriving in many poorer regions of this earth. The Tata, made in India for Indians, is not sold in the U.S. , and no American or Japanese car manufacturer even tries for that large and rapidly growing Indian market. We worry about losing markets, and about our serious imbalance of trade, but our manufacturers make not the slightest effort to open new markets either here or in other countries. "Small car, small profit" still defines today's auto market, as it has since George Romney left American Motors to become Governor of Michigan.

I recommend that we put a $1 tax on a gallon of gasoline or better 10 cents a year for ten years. That tax money should be reserved exclusively to build at least one dual fuel American car that goes more than 45 miles per gallon. I would suggest that companies give small ($1/hour?) bonuses to employees that can work at home one day a week using the Internet. Companies with cafeterias could give a free lunch to drivers who brought at least two other workers with them to work.

The big problem with transportation is that our society has almost forced us to drive more. All our policies favor suburban spread and more travel by car. These are hard customs to change, and the auto manufacturers resist it at every opportunity. Without planning people will oppose any moves to cut back on travel. "What's in it for me? "Proper incentives can make sure that there is something in it for me. The change will occur, even if slowly. Rome wasn't built a day!

Public transportation is an untapped resource. Our railroads have fallen into disrepair, and become primarily freight movers. Buses are pretty dull and old fashioned. Bus lines take some real planning. But with good planning, bus lines can be a great convenience for local people on short urban trips. The bus route is a two way street. You put bus lines where people live and when you do, more people move there to make use of them. I suspect it would be a rare occasion where a streetcar rail would be cost effective, but they should at least be considered. But remember, these changes should be designed to make life better, not just cleaner or more energy efficient. For about 50 years we built highways immediately next to railroads, inviting people to drive instead of taking the bus or train. That more than tripled the cost per mile of travel. Railroads can move freight for less than one third of the cost by truck on a super highway. It is not too late to capitalize on that parallel system and make the buses and trains truly competitive with the automobile. Where would New York City be without the subway, or Chicago without the El.? Boston and Washington, DC find their public transportation essential today. Toronto went to huge expense to build the first Canadian subway in 1954. They have not regretted it and have expanded it twice since then.

Making such changes require that people work together. Get together with your neighbors and townspeople and make plans for major beneficial changes in your local transportation. Whether its bicycles, walking paths, railroads, streetcars or buses, there are ways to make your transportation both more convenient and more energy efficient. Will you be the spark plug that makes that happen in your community? You can get lots of information on the Internet.

An important agenda for *"Re-Volt!"* is to get people to work together. Cooperation is the foundation of successful societies. In recent years the cell phone and the blog have hindered cooperation. It is easy to talk to each other, but more difficult to do things with each other if we live miles apart. The cell phone tends to make us more distant from our friends rather than closer to them. That is a serious loss. We are not getting things done so much as talking about them. Over 90% of cell phone talk is frivolous. What a waste!

I emphasize your involvement. But you have an additional option.

Government should not run things. But government should place limits and controls on private and corporate actions to assure that they are in the best interest of society rather than just for the corporation. Your government responds to you IF you tell it to. But governments seem to avoid their most important task—oversight. The automotive industry has long overreached its function as a provider of transportation. It is time government imposed controls on it to prevent further abuse. Here again, I emphasize that government should not run the transportation industry, but they must place protective rules and regulations on it and enforce them. You are the source of that government action. Get involved. Vote! Go to town meetings and to political rallies. They are your tools for, protecting yourself from greedy and ruthless corporations whose only objective is to increase their bottom line. Most great companies became great by creating something of value to the common man. That is as it should be. But when they lose sight of that original goal and begin to squeeze more out of it than it earns, government should be the restraining force. You can make that happen.

Chapter 22

True Alternatives

We have been investigating alternative energy. We have viewed it differently than most people. We are not replacing the current energy supply system, we are supplementing it. In fact what we are doing is creating a whole new system of energy in addition the very sophisticated and long-standing one we have. Capitalizing on neglected resources that now abound is an entire new industry--the Home Energy System. It will be in addition to the grid and the oil companies and the coal companies. It will employ millions of people worldwide. It will require small shops and energy servicemen and parts stores and consultants. It will be a huge industry, because in twenty or thirty years it will be as big as today's energy industry. But it will not have giant companies that control it like today's energy system and today's health care. This is a grass roots movement which will struggle to create standardized parts and procedures. There is an enormous amount of work to be done by thousands of people. It will grow rapidly and probably somewhat chaotically at first, but everyone will ultimately use and benefit from all that effort. Even the utility companies will benefit, because they will not be replaced but facilitated in what they do best while delegating the things they don't do well to others who do. Above all, it epitomizes the idea of alternatives. If two or more systems run parallel, users can switch from one to the other when a crisis or a shortage occurs. Isn't that what alternatives are? We don't want to replace one vulnerable system with another, but to create options to use any of them as needed. It also keeps them all on their toes to meet the competition.

Consider the advent of television. It travels over the same airwaves that radio does, but it did not replace radio. It expanded it. The current energy supply organizations like Con Ed, Exxon Mobil and West Virginia Coal Co. will still run at full capacity. Your house will not eliminate electric wiring or the oil burner. Before I went to kindergarten there was not even a radio. We got one of the first radios on our block when I was in kindergarten. I bought my first television set when my son was born. Your children will hopefully get their first wind generator or solar panel on their house when they are in grade school. Life moves on. I don't know what your grandchildren will find that doesn't even exist today. But it won't replace what we have today. It will add to it.

As we learned in the chapter on opportunity, alternate energy is probably the biggest opportunity in two hundred years. We are going to build one of the largest industries in the world in about three decades. Recognize what that means. All those jobs we sent overseas will be replaced by alternate energy enterprises. You will be involved. Here is your chance to find an alternate energy niche and start a new business. Employ your friends and neighbors. Work with schools and universities. Form town action committees. But start now! It will take some time, and the faster we start, the faster we will learn. Remember, we have to create 13 million jobs to replace those we gave away. The world's largest industry can employ 13 million people.

I have hesitated to talk about the dramatic downside, but there is one. If we don't take action, you can expect energy prices to double or triple within ten years. There will be more Iran's and more North Koreas. There will be more international Exxon Mobils that can charge whatever they want because there is no alternative. Our atmosphere will contain 500 ppm of CO_2 and there will be no lobsters or shellfish. Respiratory illnesses will increase dramatically. The polar bears will be gone, and Fort Lauderdale will be under water. Unemployment will remain high and our GDP will continue to decline. This is no casual thing. Energy will control our lives if we don't take action to control it. That includes you!

Jeremy Gorman

In 1930 electricity cost about 4 cents per KWH. 70 years later, despite enormous increases in efficiency and many creative improvements in delivery systems it is about 14 cents per KWH. (Don't you wish that the dollar was diluted only by three and one half times in that same period? In 1930 a hot dog cost a nickel. In electricity inflated dollars, it would cost 18 cents today.) In 2030 electricity will cost over 50 cents per KWH if we don't take action.

The focus of the industry was to squeeze all they could out of the technology they had. Why dig for gold in your neighbor's yard when you have plenty in your own back yard? To do that, they did everything they could to control the sources of energy. They discouraged wind and solar and favored huge dams, because they were subsidized by Uncle Sam, and could turn a profit. It took over 25 years to pay for Hoover Dam with electricity it generated. (Except for inflation, it still would not be paid for.) Make no mistake; they will fight any outside innovation that threatens their dominance of energy. They are not bad guys, although they have had their share of bad guys. They are just protecting their own interests vigorously. They are very powerful politically. They have well-paid lobbyists who oppose your right to get energy they cannot control. They are global companies. They have no loyalty to the country that bred them. They will move to another country if they can't get their way here. So use your own creativity and make tomorrow's world an energy driven society that does not cost an arm and a leg, and won't destroy our 65 million year-old atmosphere. Together we can do it. I hate to say so, but we have to do it.

Energy comes in many forms. This book not only discusses hundreds of sources of energy, it deals with another kind of energy--your personal energy. Many of the things in this book you have heard before, and some are repeated often. But there is one energy that dominates the message of this book. Your internal energy. How can *"Re-Volt!"* spur you to get up and do something which you already know is going to have to be done sooner or later? In some instances later won't do. If the polar bears go extinct, you cannot bring them back. Saving them will give you a sense of pride. Man has not only seen, but has caused the extinction of at least 200 species in the past century. Would you feel proud if you could stop that? You need to see a reward for your effort. It should be fun. Working together is uplifting to all the participants. You will learn fascinating things you never knew about energy, about your neighbors or committee members and your own community. You will probably learn a lot about you! Working together is fun, and is rewarding. All it takes is to get started on one project. Once you start, it will be almost addictive. Your enthusiasm and your commitment will grow together. Pick one of the projects in this book, and start work on it. Get your friends and neighbors together and make a difference. Get on the Internet, or tweet your friends, and find out who is doing what. Before you start, it will seem daunting, but once you get involved, it will become easier and more rewarding. If we create a true grass-roots energy campaign, it will be completed in remarkably short time. If we don't, it will take a century or, worst case, it will never happen. Remember, the thing that sets man apart from all other creatures is his ability to cooperate. Try it--you'll like it!

Chapter 23

Alternate is Supplemental

As we create this new world of energy, we will create jobs and opportunities for millions of people. The current energy supply organizations like Big Oil, Big Volt, and Big Coal will still run at full capacity. Your house will not eliminate electric wiring or the oil-fired boiler. Your children may get their first wind generator on their house when they are in grade school. Life moves on. I don't know what your grandchildren will find that doesn't even exist today. It shouldn't replace what we have today. It should add to it.

One of the big mistakes we made in the last thirty years is the shipping of our jobs overseas. Industry, without government oversight, sent their work overseas to capitalize on low cost labor in undeveloped nations. We didn't send our jobs over to England, or to Germany, but to Indonesia, China and India. That was bad enough, but industry made no effort to grow new replacement markets for those jobs lost. Energy was clearly one immense opportunity that we completely neglected. We got a warning in 1972 when OPEC tried to increase the price of oil to squeeze more money from the world's largest economy. After a brief success, we regained a stable oil price and promptly forgot all our conservation and alternate energy efforts and took the easy route--more reliance on oil. So our oil imports increased from 45% to about 70%. So OPEC did it again in 2008 and almost broke the bank.

Make no mistake; they will do it again and again. If OPEC doesn't, someone else will. I hope we have learned this time. Energy, as I have often repeated, is the largest industry in the world. It is growing. That presents an enormous opportunity for the creative American worker and our superb University system of education. There are hundreds of alternate resources of energy that are essentially untapped. All those jobs that we sent overseas in our misguided effort to capitalize on low wages elsewhere could and should have been replaced with jobs in energy and health care. But industry went haywire. They let the focus on the bottom line cloud their vision of our future. Don't forget, Exxon Mobil, or Con Ed can operate just as well in China as they can in the U.S. and the American worker can be completely abandoned in favor of 20 cents per hour Chinese workers. There is no shortage of them. Please note, that in 2009, Exxon Mobil earned over $12 billion but paid no U.S. income tax--their money was reportedly earned overseas where there are no U.S. Taxes. They are already becoming a global corporation. When will they move their headquarters to Bangladesh?

It is not too late. The opportunity is still there, but the competition has not stood by idly. Alternate energy is much more evolved in Western Europe and Asia than it is here. That is because of our neglect, not because of their special ingenuity. But they are learning. We allowed industry to control American energy, and industry sought only to increase their bottom line. They think internationally. They don't care if their income comes from the U.S. or from Indonesia, as long as it increases. But we gave away the earning power of the creative American worker and sent him to menial jobs at half the wages he earned before. Our GDP has stagnated, or, in inflation adjusted dollars has declined. Why? Because we have neglected one of the greatest opportunities in all history:creating a new alternate energy system that is diverse and dynamic and can support our un- and under-employed workers and once again make us the driving economic force of the planet. Our GDP can easily increase by 30%, and the big national debt we are accumulating will decline and become insignificant. We can become an exporter of energy systems and tools, and eliminate the growing imbalance of trade. Isn't that what we want? The growing trend to become international should not be at the expense of the American worker, but to his advantage as well as that of the rest of the world. Our real job is to increase the whole pie, not just to increase industry's share of the existing pie. Our industry has missed that point altogether. Let us not follow that blunder.

So crank up your ideas and start a small energy business. Perhaps you will make windmills, or solar panels. Perhaps you will learn how to make geothermal energy cost competitive. Maybe you can even figure out how to make thermoelectric power useful. Can we learn to make economical use of hydrogen? Whatever it is, join with others and create new ways to energize this nation and this world. We are crying for energy, and it is everywhere to be had. Perhaps you can figure out a way to make us all better off!

The End

www.ingramcontent.com/pod-product-compliance
Lightning Source LLC
Chambersburg PA
CBHW031054180526
45163CB00002BA/837